Experimental Evaluation Design for Program Improvement

Sara Miller McCune founded SAGE Publishing in 1965 to support the dissemination of usable knowledge and educate a global community. SAGE publishes more than 1000 journals and over 800 new books each year, spanning a wide range of subject areas. Our growing selection of library products includes archives, data, case studies and video. SAGE remains majority owned by our founder and after her lifetime will become owned by a charitable trust that secures the company's continued independence.

Los Angeles | London | New Delhi | Singapore | Washington DC | Melbourne

Experimental Evaluation Design for Program Improvement

Laura R. Peck

Abt Associates Inc.,
Social & Economic Policy Division

Los Angeles | London | New Delhi
Singapore | Washington DC | Melbourne

FOR INFORMATION:

SAGE Publications, Inc.
2455 Teller Road
Thousand Oaks, California 91320
E-mail: order@sagepub.com

SAGE Publications Ltd.
1 Oliver's Yard
55 City Road
London, EC1Y 1SP
United Kingdom

SAGE Publications India Pvt. Ltd.
B 1/I 1 Mohan Cooperative Industrial Area
Mathura Road, New Delhi 110 044
India

SAGE Publications Asia-Pacific Pte. Ltd.
18 Cross Street #10-10/11/12
China Square Central
Singapore 048423

Printed in the United States of America

Print ISBN: 978-1-5063-9005-5

This book is printed on acid-free paper.

Acquisitions Editor: Helen Salmon
Editorial Assistant: Megan O'Heffernan
Production Editor: Astha Jaiswal
Copy Editor: Diane DiMura
Typesetter: Hurix Digital
Proofreader: Ellen Brink
Indexer: Amy Murphy
Cover Designer: Candice Harman
Marketing Manager: Shari Countryman

19 20 21 22 23 10 9 8 7 6 5 4 3 2 1

BRIEF CONTENTS

DETAILED CONTENTS

LIST OF BOXES, FIGURES, AND TABLES

VOLUME EDITORS' INTRODUCTION

Impact evaluation is central to the practice and profession of evaluation. Emerging in the Great Society Era, the field of evaluation holds deep roots in the social experiments of large-scale demonstration programs—Campbell's utopian ideas of an Experimenting Society. Since then, the fervent search for "what works"—for establishing the impact of social programs—has taken on many different forms. From the early emphasis on experimental and quasi-experimental designs, through the later emergence of systematic reviews and meta-analysis, and onwards to the more recent and sustained push for evidence-based practice, proponents of experimental designs have succeeded in bringing attention to the central role of examining the effectiveness of social programs (however we chose to define it). There is a long and rich history of measuring impact in evaluation.

The landscape of impact evaluation designs and methods has grown and continues to grow. Innovative variants of and alternatives to traditional designs and approaches continue to emerge and gain prominence, addressing not only "what works" but also "what works, for whom, and under what circumstances" (Stern et al., 2012). For the novice (and perhaps even the seasoned) evaluator, the broadening array of designs and methods, not to mention the dizzying array of corresponding terminology, may invoke a mixed sense of methodological promise and peril, opportunity and apprehension. How can randomization be applied across multiple treatments, across multiple treatment components, and across stages of a program process? What exactly is the difference between multistage, staggered, and blended impact evaluation designs? And are there any practical and methodological considerations that one should award particular attention to when applying these designs in real-world settings?

These are but a few of the questions answered in Laura Peck's *Experimental Evaluation Design for Program Improvement*. Grounded on decades of scholarship and practical experience with real-world impact evaluation, Peck begins the book by providing a concise and accessible introduction to the "State of the Field," carefully guiding the reader through decades of developments in the experimental design tradition, including large-scale experimental designs, nudge experiments, rapid cycle evaluation, systematic reviews and the associated meta-analysis, and more recently design options for understanding impact variation across program components.

After this introduction, Peck describes a "framework for thinking about the aspects of a program that drive its impacts and how to evaluate the relative contributions of those aspects," rooted in the idea of using a well-developed program logic model to discern the most salient program comparisons to be examined in the evaluation. As Peck states, "From a clear and explicit program logic model, the evaluation logic model can also be framed to inform program operators' understanding of the essential ingredients of their programs" (p. 27). The remainder of the book is dedicated to a broad variety of experimental design options for measuring program impact, covering both traditional designs and more recent variants of these (e.g., multistage and blended designs). Bringing these designs closer to practice, an illustrative application and a set of practical lessons learned are provided. A set of hands-on principles for "good practice" concludes the book.

The present book is an important contribution to the growing landscape of impact evaluation. With her aim to identify a broader range of designs and methods that directly address causal explanation of "impacts," Peck opens new frontiers for impact evaluation. Peck directly challenges, and correctly so, the longstanding perception that experimental designs are unable to get inside the black box of how, why, and for whom social programs work. Countering this idea, Peck describes and illustrates by way of practical examples, a variety of design options that each in their own way support causal explanation of program impact. By doing so, the book significantly broadens the types of evaluation questions potentially pursued by experimental impact evaluations. As Peck opines, "using experimental evaluation designs to answer 'black box' type questions—what works, for whom, and under what circumstances—holds substantial promise" (p. 10). We agree.

Sebastian T. Lemire, Christina A. Christie, and Marvin C. Alkin
Volume Editors

REFERENCE

Stern, E., Stame, N., Mayne, J., Forss, K., Davies, R., & Befani, B. (2012). Broadening the range of designs and methods for impact evaluations. Report of a study commissioned by the Department for International Development. Working Paper 38. Accessible from here: https://www.oecd.org/derec/50399683.pdf

ABOUT THE AUTHOR

Laura R. Peck, PhD, is a principal scientist at Abt Associates and has spent her career evaluating social welfare and employment policies and programs, both in research and academic settings. A policy analyst by training, Dr. Peck specializes in innovative ways to estimate program impacts in experimental and quasi-experimental evaluations, and she applies this to many social safety net programs. Dr. Peck is currently the principal investigator, co-PI, or director of analysis for several major national evaluations for the U.S. Departments of Health and Human Services, Labor, and Housing and Urban Development and over her career has been part of more than 35 informative evaluations of nonprofit, local, state, and federal programs and policies. Peck is a co-author of a public policy textbook and is well published on program evaluation topics. Prior to her work at Abt Associates, Dr. Peck was a tenured professor at the Arizona State University School of Public Affairs and also served as the founding associate dean of the Barrett Honors College, Downtown Phoenix campus. She earned her PhD from the Wagner Graduate School at New York University.

ACKNOWLEDGMENTS

This book has been motivated by many years of scholarly and applied evaluation work. Along the way, I developed and tested my ideas in response to prodding from professors, practitioners, and policymakers; and it is my hope that articulating them in this way gives them more traction in the field. Through my graduate school training at New York University's Wagner School, I especially valued the perspectives of Howard Bloom, Jan Blustein, and Dennis Smith. While at Arizona State University's School of Public Affairs, I appreciated opportunities to engage many community agencies, who served as test beds for my graduate students as they learned about evaluation in practice.

Most recently, at Abt Associates, I am fortunate to have high-quality colleagues and projects where we have the chance not only to inform public policy but also to advance evaluation methods. I am grateful to the funders of that research (including the U.S. Departments of Labor, Health and Human Services, and Housing and Urban Development) who are incredibly supportive of advancing evaluation science: They encourage rigor and creativity and are deeply interested in opening up that black box—be it through advancing analytic approaches or design approaches. I am also grateful to have had the opportunity to write about some of these ideas in the context of my project work, for the American Evaluation Association's AEA365 blog, and for Abt Associates' *Perspectives* blog.

Evaluation is a team sport, and so the ideas in this book have evolved through teamwork over time. For example, some of the arguments, particularly regarding the justification for experiments and ethics of experimentation (in Chapter 1), stem from work with Steve Bell (and appear in our joint *JMDE* publication in 2016). The discussion of whether a control group (in Chapter 4) is needed as well as some observations of design variants draws on earlier work as well (e.g., Peck, 2015, in *JMDE*; Bell & Peck, 2016, in *NDE* #152). In addition, some of the (Appendix) discussion of the trade-offs between intent-to-treat and treatment-on-the-treated impacts and the factors that determine minimum detectable effect sizes came from joint work with Shawn Moulton, Director of Analysis, and the project team for the HUD First-Time Homebuyer Education and Counseling Demonstration.

At Abt Associates, Rebecca Jackson provided research assistance, Bry Pollack provided editorial assistance, Daniel Litwok provided critical review of a draft manuscript, and the Work in Progress Seminar offered input on final revisions. I am also appreciative of input from five anonymous reviewers for the *Evaluation in Practice Series*, and SAGE editors.

My most intellectually and personally enriching partnership—and longest-standing collaboration—is with Brad Snyder who asks the right and hard questions, including the "and then," which implies pushing further still. I also thank my parents for raising me to value questioning and my daughter for teaching me patience in answering.

I would also like to acknowledge the following reviewers for their feedback on the book:

Deven Carlson, *University of Oklahoma*

Roger A. Boothroyd, *University of South Florida*

Sebastian Galindo, *University of Florida*

Katrin Anacker, *George Mason University*

Christopher L. Atkinson, *University of West Florida*

Sharon Kingston, *Dickinson College*

Colleen M. Fisher, *University of Minnesota*

Regardt Ferreira, *Tulane University*

1 INTRODUCTION

The concepts of cause and effect are critical to the field of program evaluation. After all, establishing a causal connection between a program and its effects is at the core of what impact evaluations do. The field of program evaluation has its roots in the social work research of the settlement house movement and in the business-sector's efficiency movement, both at the turn of the 20th century. Evaluation as we know it today emerged from the Great Society Era, when large scale demonstrations tested new, sweeping interventions to improve many aspects of our social, political, and economic worlds. Specifically, it was the Elementary and Secondary Education Act of 1965 that first stipulated evaluation requirements (Hogan, 2007). Thereafter, a slew of scholarly journals launched and, to accompany them, academic programs to train people in evaluation methods. Since then scholars, practitioners and policymakers have increased their awareness of the diversity of questions that program evaluation pursues. This has coupled with a broadening range of evaluation approaches to address not only whether programs work but also *what* works, *for whom,* and *under what circumstances* (e.g., Stern et al., 2012). Program evaluation as a profession is diverse, and scholars and practitioners can be found in a wide array of settings from small, community-based nonprofits to the largest of federal agencies.

As those program administrators and policymakers seek to establish, implement, and evolve their programs and public policies, measuring the effectiveness of the programs or policies is essential to justifying ongoing funding, enacting policy changes to improve it, or terminating. In doing so, impact evaluations must isolate a program's impact from the many other possible explanations that exist for any observed difference in outcomes. How much of the improvement in outcomes (that is, the "impact") is due to the program involves estimating what would have happened in the program's absence (the "counterfactual"). As of 2019, we are amid an era of "evidence-based" policy-making, which implies that the results of evaluation research inform what we choose to implement, how we choose to improve, and whether we terminate certain public and nonprofit programs and policies.

Experimentally designed evaluations—those that randomize to treatment and control groups—offer a convincing means for establishing a causal connection between a program and its effects. Over the last roughly 3 decades, experimental evaluations have been growing substantially in numbers and diversity of their application. For example, Greenberg and Shroder's 2004 *Digest of Social Experiments* counted 293 such evaluations since the beginning of their use to study public policy in the 1970s. The *Randomized Social Experiments eJournal* that replaced the *Digest* beginning in 2007 identifies additional thousands of experiments since then.

The past few decades have shown that experimental evaluations are feasible in a wide variety of settings. The field has gotten quite good at executing experiments that aim to answer questions about average impacts of policies and programs. Over this same time period there has been increased awareness of a broad range of cause-and-effect questions that evaluation research examines and corresponding methodological innovation and creativity to meet increased demand from the field. That said, experimental evaluations have been subject to criticism, for a variety of reasons (e.g., Bell & Peck, 2015).

The main criticism that compels this book is that experimental evaluations are not suited to disaggregating program impacts in ways that connect to program implementation or practice. That is, experiments have earned a reputation for being a relatively blunt tool, where program implementation details are a "black box." The complexity, implementation, and nuance of a program itself tends to be overlooked when an evaluation produces a single number (the "impact") to represent the program's effectiveness.

BOX 1.1
DEFINITION AND ORIGINS OF THE TERM "BLACK BOX" IN PROGRAM EVALUATION

In the field of program evaluation, "black box" refers to how some impact evaluations are perceived to consider the program and its implementation. It is possible to evaluate the impact of a program without knowing much at all about what the program is. In that circumstance, the program itself is considered a black box, an unknown.

Perhaps the first published reference to black box appeared in a 1993 Institute for Research on Poverty discussion paper, "Prying the Lid from the Black Box" by David Greenberg, Robert Meyer, and Michael Wiseman (although two of these authors credit Larry Orr for using the *black box* term before then). This paper seems to have evolved and was published in 1994 as "Multisite Employment and Training Program Evaluation: A Tale of Three Studies" by the same trio, with follow-up papers in the decade that followed (e.g., Greenberg Meyer, Michalopoulos, & Wiseman, 2003).

In the ensuing two decades, the term—as in *getting inside the black box*—has become associated with the idea of understanding the details of a program's operations. A special section of the *American Journal of Evaluation* (volume 36, issue 4) titled 'Unpacking the "Black Box" of Social Programs and Policies' was dedicated to the methods; and three chapters of the 2016 *New Directions for Evaluation* (issue 152) considered "Inside the Black Box" evaluation designs and analyses.

Indeed, recent years have seen policymakers and funders—in government, private, and foundation sectors—desiring to learn more from their evaluations of health, education, and social programs. Although the ability to establish a program's causal impact is an important contribution, it may be insufficient for those who immediately want to know what explains that treatment effect: Was the program effective primarily because of its quality case management? Did its use of technology in interacting with its participants drive impacts? Or are both aspects of the program essential to its effectiveness?

To answer these types of additional research questions about the key ingredients of an intervention's success with the same degree of rigor requires a new perspective on the use of experimentals in practice. This book considers a range of impact evaluation questions, most importantly those questions that focus on the impact of specific aspects of a program. It explores how a variety of experimental evaluation design options can provide the answers to these questions and suggests opportunities for experiments to be applied in more varied settings and focused on program improvement efforts.

THE STATE OF THE FIELD

The field of program evaluation is large and diverse. Considering the membership and organizational structure of the U.S.-based American Evaluation Association (AEA)—the field's main professional organization—the evaluation field covers a wide variety of topical, population-related, theoretical, contextual, and methodological areas. For example, the kinds of topics that AEA members focus on—as defined by the association's sections, or Topical Interest Groups (TIGs), as they are called—include education, health, human services, crime and justice, emergency management, the environment, and community psychology. As of this writing, there are 59 TIGs in operation. The kinds of population-related interests cover youth; feminist issues; indigenous peoples; lesbian, gay, bisexual and transgendered people; Latinos/as; and multiethnic issues. The foundational, theoretical, or epistemological perspectives that interest AEA members include theories of evaluation, democracy and governance, translational research, research on evaluation, evaluation use, organizational learning, and data visualization. The contexts within which AEA members consider their work involve nonprofits and foundations, international and cross-cultural entities and systems, teaching evaluation, business and management, arts and cultural organizations, government, internal evaluation settings, and independent consultancies. Finally, the methodologies considered among AEA members include collaborative, participatory, and empowerment; qualitative; mixed methods; quantitative; program-theory based; needs assessment; systems change; cost-benefit and effectiveness; cluster, multisite, and multilevel; network analysis; and experimental design and analytic methods, among others. Given this diversity, it is impossible to classify the entire field of program evaluation neatly into just a few boxes. The literature regarding any one of these topics is vast, and the intersections across dimensions of the field imply additional complexity.

What this book aims to do is focus on one particular methodology: that of experimental evaluations. Within that area, it focuses further on designs to address the more nuanced questions what about a program drives its impacts. The book describes the basic analytic approach to estimating treatment effects, leaving full analytic methods to other texts that can provide the needed deeper dive.

Across the field, alternative taxonomies exist for classifying evaluation approaches. For example, Stern et al. (2012) identify five types of impact evaluations: experimental, statistical, theory based, case based, and participatory. The focus of this book is the first. Within the subset of the evaluation field that uses randomized experiments, there are several kinds of evaluation models, which I classify here as (1) **large-scale experiments**, (2) **nudge** or **opportunistic experiments**, (3) **rapid-cycle evaluation**, and (4) **meta-analysis** and **systematic reviews**.

Large-Scale Experiments

Perhaps the most commonly thought of experiments are what I will refer to as "large-scale" impact studies, usually government-funded evaluations. These tend to be evaluations of federal or state policies and programs. Many are demonstrations, where a new program or policy is rolled out and evaluated. For example, beginning in the 1990s, the U.S. Department of Housing and Urban Development's Moving to Opportunity Fair Housing Demonstration (MTO) tested the effectiveness of a completely new policy: that of providing people with housing subsidies in the form of vouchers under the condition that they move to a low poverty neighborhood (existing policy did not impose the neighborhood poverty requirement).

Alternatively, large-scale federal evaluations can be reforms of existing programs, attempts to improve incrementally upon the status quo. For instance, a slew of welfare reform efforts in the 1980s and 1990s tweaked aspects of existing policy, such as changing the tax rate on earnings and its relationship to cash transfer benefit amounts, or changing the amount in assets (such as a vehicle's value) that a person could have while maintaining eligibility for assistance. These large-scale experiments usually consider broad and long-term implications of policy change, and, as such, take a fair amount of time to plan, implement, and generate results.

This slower process of planning and implementing a large-scale study, and affording the time needed to observe results, is also usually commensurate with the importance of the policy decisions: Even small effects of changing the tax rate on earnings for welfare recipients can result in large savings (or costs) nationally. Although we might hope for—or seek out—policy changes that have large impacts, substantial, useful policy learning has come from this class of experimental evaluations (e.g., Gueron & Rolston, 2013; Haskins & Margolis, 2014). For example, the experimentation that focused on reforming the U.S. cash public assistance program was incremental in its influence. That program's evaluation—Aid to Dependent Children (ADC) and Aid to Families with Dependent Children (AFDC) from 1935 until 1996 and Temporary Assistance for Needy Families (TANF) since then—amassed evidence that informed many policy changes. Evidence persuaded policymakers to change various aspects of the program's rules, emphasize a work focus rather than an education one, and end the program's entitlement.

Nudge or Opportunistic Experiments

In recent years, an insurgence of "opportunistic" or "nudge" experiments has arisen. An "opportunistic" experiment is one that takes advantage of a given opportunity. When a program has plans to change—for funding or administrative reasons—the evaluation can take advantage of that and configure a way to learn about the effects of that planned change. A "nudge" experiment tends to focus on behavioral insights or administrative systems changes that can be randomized in order to improve program efficiency. Both opportunistic and nudge experiments tend to involve relatively small changes—such as to communications or program enrollment or compliance processes—but they may apply to large populations such that even a small change can result in meaningful savings or benefits. For example, in the Fall of 2015, the Obama administration established the White House Social and Behavioral Sciences Team (SBST) to improve administrative efficiency and embed experimentation across the bureaucracy, creating a culture of learning and capitalizing on opportunities to improve government function.

The SBST 2016 Annual Report highlights 20 completed experiments that illustrate how tweaking programs' eligibility and processes can expand access, enrollment, and related favorable outcomes. For instance, a test of automatic enrollment into retirement savings among military service members boosted enrollment by 8.3 percentage points from a low of 44% to over 52%, a start at bringing the savings rate closer to the 87% among civilian federal employees. Similarly, waiving the application for some children into the National School Breakfast and Lunch program increased enrollment, thereby enhancing access to food among vulnerable children. Both of these efforts were tested via an experimental evaluation design, which randomized who had access to the new policy so that the difference between the new regime's outcomes and the outcomes of the status quo could be interpreted as the causal result of the new policy. In both cases, these were relatively small administrative changes that took little effort to implement; they could be implemented across a large system, implying the potential for meaningful benefits in the aggregate.

Rapid-Cycle Evaluation

Rapid-cycle evaluation is another relatively recent development within the broader field of program evaluation. In part because of its nascency, it is not yet fully or definitively defined. Some scholars assert that rapid-cycle evaluation must be experimental in nature, whereas others define it as any quick turnaround evaluation activity that provides feedback to ongoing program development and improvement. Regardless, rapid-cycle evaluations that use an experimental evaluation design are relevant to this book. In order to be quick-turnaround, these evaluations tend to involve questions similar to those asked by nudge or opportunistic experiments and outcomes that can be measured in the short term and still be meaningful. Furthermore, the data that inform impact analyses for rapid-cycle evaluations tend to come from administrative sources that are already in existence and therefore quicker to collect and analyze than would be the case for survey or other, new primary data.

Meta-Analysis and Systematic Reviews

The fourth set of evaluation research relevant to experiments involves **meta-analysis**, including **tiered-evidence reviews**. Meta-analysis involves quantitatively aggregating other evaluation results in order to ascertain, across studies, the extent and magnitude of program impacts observed in the existing literature. These analyses tend to prioritize larger and more rigorous studies, down-weighting results that are based on small samples or that use designs that do not meet criteria for establishing a causal connection between a program and change in outcomes. Indeed, some meta-analyses use only evidence that comes from experimentally designed evaluations. Likewise, **evidence reviews**—such as those provided by the What Works Clearinghouse (WWC) of the U.S. Department of Education—give their highest rating to evidence that comes from experiments. Because of this, I classify meta-analyses as a type of research that is relevant to experimentally designed evaluations.

Getting Inside the Black Box

Across these four main categories of experimental evaluation, there has been substantial activity regarding moving beyond estimating the **average treatment effect** to understand more about how impacts vary across a variety of dimensions. For example, how do treatment effects vary across subgroups of interest? What are the **mediators** of treatment effects? How do treatment effects vary along dimensions of program implementation features or the fidelity of implementation to **program theory**? Most efforts to move beyond estimating the average treatment effect involve *data analytic* strategies rather than *evaluation design* strategies. These analytic strategies have been advanced in order to expose what is inside the "black box."

As noted in Box 1.1, the black box refers to the program as implemented, which can be somewhat of a mystery in impact evaluations: We know that the impact was *this*, but we have little idea *what* caused the impact. In order to expose what is inside the black box, impact evaluations often are paired with **implementation evaluation**. The latter provides the detail needed to understand the program's operations. That detail is helpful descriptively: It allows the user of the evaluation to associate the impact with some details of the program from which it arose. The way I have described this is at an aggregate level: The program's average impact represents what the program as a whole did or offered. Commonly, a program is not a single thing: It can vary by setting, in terms of the population it serves, by design elements, by various implementation features, and also over time. The changing nature of interventions in practice demands that evaluation also account for that complexity.[1]

Within the field of program evaluation, the concept of impact variation has gained traction in recent years. The program's average impact is one metric by which to judge the program's worth, but that impact is likely to vary along multiple dimensions. For example, it can vary for distinct subgroups of participants. It might also vary depending on program design or implementation: Programs that offer X and Y might be more effective than those offering only X; programs where frontline staff have greater experience or where the program manager is an especially dynamic leader might be more effective than those without. These observations about what makes up a program and how it is implemented have become increasingly important as potential drivers of impact.

Accordingly, the field has expanded the way it thinks about impacts, to be increasingly interested in *impact variation*. Assessments of how impacts vary—what works, for whom, and under what circumstances—are currently an important topic within the field. The field has expanded its toolkit of *analytic strategies* for understanding impact variation to addressing "what works" questions, this book will focus on *design options* for examining impact variation.[2]

THE ETHICS OF EXPERIMENTATION

Prior research and commentary considers whether it is ethical to randomize access to government and nonprofit services (e.g., Bell & Peck, 2016). Are those who "lose the lottery" and are randomized into the control group disadvantaged in some way (and is that disadvantage actually unfair or unethical)? Randomizing who gets served is just one way to ration access to a funding-constrained program. I argue that giving all deserving applicants an equal chance through a lottery is the fairest, most ethical way to proceed when not all can be served. I assert that is it unfair and unethical to hand pick

---

[1] In Peck (2015), I explicitly discuss "programmatic complexity" and "temporal complexity" as key factors that suggest specific evaluation approaches, both in design and analysis.

[2] For a useful treatment of the relevant analytic strategies—including an applied illustration using the Moving to Opportunity (MTO) demonstration—I refer the reader to Chapter 7 in *New Directions for Evaluation* #152 (Peck, 2016).

applicants to serve because that selection can involve prejudices that result in unequal treatment of individuals along racial, ethnic, nationality, age, sex, or orientation lines. Even a first-come, first-serve process can advantage some groups of individuals over others. Random assignment such as a lottery can ensure that no insidious biases enter the equation of who is served.

Furthermore, program staff can be wonderfully creative in blending local procedures with randomization in order to ensure that they are serving their target populations while preserving the experiment's integrity. For example, the U.S. Department of Health and Human Services's Family and Youth Services Bureau (FYSB) is operating an evaluation of a homeless youth program called the Transitional Living Program (Walker, Copson, de Sousa, McCall, & Santucci, 2019; U.S. Department of Health and Human Services (DHHS), n.d.a). The evaluation worked with program staff to help them use their existing needs-assessment tools to prioritize youth for the program in conjunction with a randomization process that considers those preferences: It is a win-win arrangement. Related scholarship has established procedures for embedding preferences within randomization (Olsen, Bell, & Nichols, 2017), ensuring the technical aspects of the approach as well as mitigating program concerns about ethics.

Even if control group members either are perceived to be or actually are disadvantaged, random assignment still might not be unethical (Blustein, 2005). For example, society benefits from accurate information about program effectiveness and, accordingly, research may be justified in allowing some citizens to be temporarily disadvantaged in order to gather information to achieve wider benefits for many (e.g., Slavin, 2013). Society regularly disadvantages individuals based on government policy decisions undertaken for nonresearch reasons. An example that disadvantages some people daily is that of high-occupancy vehicle (HOV) lanes: they disadvantage solo commuters to the benefit of carpoolers. Unlike an evaluation's control group exclusions, those policy decisions (such as establishing HOV lanes) are permanent not temporary.

In an example from the private sector, Meyer (2015) argues that managers who engage in **A/B testing**—where staff are subjected to alternative policies—without the consent of their employees operate more ethically than those who implement a policy change without evidence to support that change. Indeed, the latter seems "more likely to exploit her position of power over users or employees, to treat them as mere means to the corporation's ends, and to deprive them of information necessary for them to make a considered judgment about what is in their best interests" (Meyer, 2015, p. 279).

Moreover, in a world of scarce resources, I argue that it is unethical to continue to operate ineffective programs. Resources should be directed toward program improvement (or in some cases termination) when evidence suggests that a program is not generating desired impacts. From this alternative perspective, it is unethical *not* to use rigorous impact evaluation to provide strong evidence to guide spending decisions.

It is worth noting that policy experiments are in widespread use, signaling that society has already judged them to be ethically acceptable. Of course it is always essential to ensure the ethics of evaluation research, not only in terms of design but also in terms of treatment of research participants. Moreover, I acknowledge that there are instances where it is clearly unethical—in part because it may also be illegal—to randomize an individual out of a program. For example, entitlement programs in the U.S. *entitle* people to a benefit, and that entitlement cannot and should not be denied, even for what might be valuable research reasons. That does not imply, however, that we cannot or should not continue to learn about the effectiveness of entitlement programs. Instead, the kinds of questions that we ask about them are different from "Do they work?" That is, the focus is less on the overall, average treatment effects and more about the impact variation that arises from variation in program design or implementation. For instance, we might be interested to know *what level* of assistance is most effective for achieving certain goals. A recent example of this involves the U.S. Department of Agriculture's extension of children's food assistance into the summer. The Summer Electronic Benefits Transfer for Children (SEBTC) Demonstration that replaced *no* summer cash/near-cash assistance with a stipend for $30 or $60 per month is indeed an ethical (and creative) way to ascertain whether such assistance reduces hunger among vulnerable children when school is out of session (Collins et al., 2016; Klerman, Wolf, Collins, Bell, & Briefel, 2017).

This leads to my final point about ethics. Much of the general concern is about randomizing individuals into a "no services" control group. But, as the remainder of this book elaborates, conceiving the control group that way is unnecessary. Increasingly, experimental evaluation designs are being used to compare alternative treatments to one another rather than compare some stand-alone treatment to nothing. As such, concerns about ethics are much assuaged. As we try to figure out whether Program A is better or worse than Program B, or whether a program should be configured *this* way or *that* way, eligible individuals get access to something. When research shows which "something" is the better option, then all individuals can begin to be served through that better program option.

WHAT THIS BOOK COVERS

This book considers a range of experimental evaluation designs, highlighting their flexibility to accommodate a range of applied questions of interest to program managers. These questions about impact variation—what drives successful programs—have tended to be outside the purview of experimental evaluations. Historically, they have been under the purview of nonexperimental approaches to impact evaluation, including theory-driven evaluation, case-based designs, and other, descriptive or correlational, analytical strategies. It is my contention that experimental evaluation designs, counter to common belief among many an evaluator, can actually be used to address what works, for whom, and under what circumstances.

It is my hope that the designs discussed will motivate their greater use for program improvement for the betterment of humankind.

- Why a focus on *experimental* evaluation? I focus on *experimental* evaluation because of its relative importance to funders, its ability to establish causal evidence, and its increasing flexibility to answer questions addressing more than the average treatment effect.

- Why a focus on experimental evaluation *designs*? I focus on experimental evaluation *designs* because (1) alternative, nonexperimental designs are covered in other texts, and (2) many analytic strategies aimed at uncovering insights about "black box" mechanisms necessitate specialized analytic training that is beyond the scope of this book.

- Why *not* a focus on nonexperimental designs and analysis strategies? There is substantial, active research in the "design replication" (or "within-study comparison") literature that considers the conditions under which nonexperimental designs can produce the same results as an experimental evaluation. As with advanced analytic strategies, is it beyond the scope of this book to offer details—let alone a primer—on the many, varied nonexperimental evaluation designs. Suffice it to say that those designs exist and are the subjects of other books.

Using experimental evaluation designs to answer "black box" type questions—what works, for whom, and under what circumstances—holds substantial promise. Making a shift from thinking about a denied control group toward thinking about comparative and enhanced treatments opens opportunities for connecting experimental evaluation designs to the practice of program management and evidence-based improvement efforts.

The book is organized as follows: After this Introduction, Chapter 2 suggests a conceptual framework, building from the well-known program logic model and extending that to an evaluation logic model. Chapter 3 offers an introduction to the two-group experimental evaluation design. As the center of the book, Chapter 4 considers variants on experimental evaluation design that are poised to answer questions about program improvement. Chapter 5 concludes by discussing some practical considerations and identifying some principles for putting experimental evaluation into practice. Finally, an Appendix provides basic instruction in doing the math needed to generate impact estimates associated with various designs. When randomization is used, the math can be quite simple. The Appendix also addresses the relationship between sample size impact magnitude. Each of the chapters ends with two common sections: Questions and Exercises, and Resources for Additional Learning.

QUESTIONS AND EXERCISES

1. Identify an applied experimental evaluation that fits each type of experimental evaluation model: large-scale, nudge/opportunistic, rapid-cycle, and meta-analysis or systematic review.

2. Discuss: To what extent do you agree or disagree with each of the arguments about the ethics of experimentation?

RESOURCES FOR ADDITIONAL LEARNING

- Rigorous Evaluations and Evidence-Based Policy and Innovation Initiative (of the Laura and John Arnold Foundation): https://www.arnoldventures.org/work/evidence-based-policy

- Government Innovator blog: http://govinnovator.com/

- U.S. Office of Management and Budget's Evidence Team: https://obama whitehouse.archives.gov/omb/evidence; https://www.whitehouse.gov/omb/information-for-agencies/evidence-and-evaluation/

- Social and Behavioral Sciences Team (SBST): https://sbst.gov/

2 CONCEPTUAL FRAMEWORK

From Program Logic Model to Evaluation Logic Model

This chapter presents a framework for thinking about the aspects of a program that drive its impacts and how to evaluate the relative contributions of those aspects. I present two perspectives—a program operator's perspective and an evaluator's perspective—the **program logic model** and the evaluation logic model.

Program operators are keenly familiar with the concept of the logic model because funders tend to require it, and because it is the underpinning to most evaluation work.[1] In recent years, scholars and practitioners alike have tended to use the term *logic model* interchangeably with a theory of change. They are not the same. A **logic model** articulates a program's **inputs**, **activities**, output, and **outcomes**; a theory of change incorporates information about **plausible rival explanations** for a change in outcomes that might occur. A **theory of change** includes information from outside of the program that influences the program and should be accounted for in understanding how the program is hypothesized to work. A theory of change can be embedded into a full-blown approach to evaluation where a fully articulated theory of change is used as a kind of qualitative counterfactual against which change is assessed (e.g., Connell & Kubisch, 1998; Connell Kubisch, Schorr, & Weiss 1995).

Large nonprofit funders, such as the United Way, and major government funders such as the Centers for Disease Control and Prevention (CDC) require that their grantees undertake a logic model exercise. Doing so helps make explicit how a program is expected to operate and generate its results. Whether required by funders or not, local programs should engage in creating and using logic models as a means to articulate how their programs are assumed to work. From there, the logic model can serve as a means for framing evaluation questions. In current practice, however, logic models are rarely actually used. To have utility for both evaluation and for program learning, the program logic model must be actively examined as part of the ongoing process of program assessment.

In order to make optimal use of the material covered in Chapters 3 and 4, a program's logic model must be clear and explicit, including assumptions about the program's mediators and how certain activities are hypothesized to lead to certain outputs and outcomes. Then this clear and explicit logic model must be linked to the associated evaluation logic model.

---

[1] The concept of a program logic model has several synonyms, including **conceptual framework**, **program logic**, **program flow**, **program theory**, theory of change, **causal model**, **results chain**, and **intervention logic**.

Each of the program and evaluation perspectives on the logic model is discussed in turn.

PROGRAM LOGIC MODEL

A program's logic model reflects how, in practice, the program is expected to convert its inputs into activities, which result in outputs that generate desired outcomes. These elements—inputs, activities, outputs, and outcomes—are the four key parts of a standard logic model. The W. K. Kellogg Foundation offers a useful guide for how to undertake the development of a program logic model (W. K. Kellogg, n.d.). In order for a program to articulate how it expects to *change* the outcomes identified on the right-hand side of its logic model, the four elements of the logic model must be identified and their relationships stated.

Figure 2.1 presents an illustration of a program logic model for a job training program. Program inputs include funding and the program's design, administration and implementation plans, and participants or clients. The activities for the program include intake, enrollment, and assessment; education and training; support services; and employer connections. These activities are expected to generate the program outputs of individuals who have completed education or training, developed skills, and have a career plan in place. In turn, these outputs (results of the program activity) are expected to lead to outcomes related to educational progress, including credential attainment, and employment in quality jobs with family-sustaining earnings.

In order to articulate the details of the program logic model, we can think about the program targets as they experience the program. In the example above, an individual person seeking job training is the target and identified as a "client" among the inputs. In turn, that client participates in the activities offered, and the subsequent output is the individuals who have completed the training course. As such, when program administrators think about their outputs, they can articulate those outputs in terms of the numbers of clients who complete trainings offered. Finally, it is those

FIGURE 2.1 ■ Illustrative Job Training Program Logic Model

targets whose outcomes are measured: the educational attainment that the participants achieve, the quality of jobs they secure, and their well-being.

An important point about the program logic model is that the outcomes included
on the right-hand side are constructs that the participants experience regardless of
whether there is a program in place to influence those outcomes. For example, people
will be employed or not, whether a program influences their employment. Although
common logic model exercises encourage programs to identify the outcomes that
they expect to achieve (e.g., being employed), the simple process of identifying those
outcomes does not necessarily mean that the program has caused or will cause *change*
in those outcomes. The logic model is more of a planning and implementation tool
than an evaluation tool, although it provides a useful foundation for evaluation. To
follow the outcomes box on the right-hand side of the logic model requires some
evaluation strategy (including the designs discussed in this book) that will assist a
program in identifying its contribution to *changing* (ideally, *improving*) those outcomes. This *change* in outcomes is the program's **impact**.

Weiss, Bloom, and Brock (2013) refer to program outputs as "intermediate outcomes" and program outcomes as "target outcomes." This semantic distinction helps
distinguish between what a program simply produces (its outputs or intermediate
outcomes) via its activities and what it aims to achieve (its target outcomes). This
semantic distinction will also be helpful in moving from the program logic model to
an evaluation logic model.

Important for this book's charge to inform program improvement through use
of experimental evaluation is the use of the program logic model to identify certain
aspects of a program that are hypothesized to be especially important to generating
impacts. This means that the box that is labeled "activities" in Figure 2.1 must be
more fully elaborated, including how each of those several activities associates with
specific outputs and outcomes. For example, considering the program activities box,
is there one or another of those activities that is most important for the program to
have in place in order to achieve its desired outcomes? Or must all of the activities
be in place for the program be effective? From the program perspective, articulating
these elements of the logic model is a necessary first step. Once those elements and
the logic of their relative roles are identified, then the program is poised to engage in
evaluation to further assess the relationships between and among specific program
activities, their outputs, and subsequent outcomes. For example, rather than group-
ing all elements of the activities box together and having a single arrow implying a
relationship to an outcomes box, *each* element of the activities box should be linked
to *each* specific relevant output that it is expected to generate.

Consider again the illustrative job training logic model in Figure 2.1. The activities box contains the following: intake, enrollment; assessments; education, training;
support services; and employer connections. To make each of these more explicit,
the next step should be to articulate how each of the activities relates to its expected

outcomes. For example, a program might let an intake worker choose which assessment tool to use given the characteristics of the person seeking training. The U.S. Department of Labor's CareerOneStop (www.careeronestop.org) offers an "Interest Assessment," a "Skills Matcher Assessment," and an "Interest Profile." Each of these has a different focus and goal. The Interest Assessment and the Interest Profile both aim to help job seekers match their interests to careers, whereas the Skills Matcher Assessment helps match one's skills to careers that use them. One of these assessments might help job seekers find and subsequently complete training paths that align with their interests or skills. Training then associates with better job placement and subsequent labor market benefits.

Next, consider the training itself. In the general logic model, the training is simply listed as "training." A more explicit logic model would identify the various characteristics of the training that are hypothesized to link to improved labor market outcomes for participants. For example, the training might involve an applied, hands-on module, a workplace skills component, or the opportunity to engage in an internship.

Finally, consider the support services listed in the activities box. An explicit logic model would identify the specific services available, such as on-site child care providers, transportation subsidies, or one-on-one intensive case management. Each of these program activities—assessments, training content, support services—might have its own independent influence on outcomes, or it might be that the combination of all of them is what is essential for achieving change.

From a clear and explicit logic model, the evaluation logic leverages that information, identifies the evaluation questions, provides a setting for identifying a **counterfactual**, and then estimates the extent to which a change in outcomes (the impact) arises because of the program's efforts.

EVALUATION LOGIC MODEL

An evaluation of the average treatment effect or of impact variation starts from the program's logic model and builds from there. Once the relationships between and among program activities, outputs, and outcomes have been articulated, the evaluation perspective frames what is necessary in order to identify whether a *change in outcomes* (i.e., impact) can be credited to the program. To do so also requires ruling out plausible rival explanations for that change (Chapter 1 defined these "plausible rival explanations" as **threats to internal validity**). The evaluation perspective on the logic model involves identifying the *contrasts* that permit us to know the source and implications of program impacts and variation in impacts. Weiss et al. (2013) articulated this as a conceptual framework for evaluating variation in program impacts. I recast that work here, with explicit relationship to the program logic, to connect it to the evaluation design options that this book explores. That is, the evaluation logic is helpful for conceptualizing an overall impact analysis, and it can also be extended to identify the relationships between certain program activities and their contributions to program impacts. From a clear and explicit program logic model, the evaluation logic model can also be framed to inform program operators' understanding of the essential ingredients of their programs.

The evaluation logic model involves positioning two logic models adjacent to each other: One is the program logic model, and the other represents the counterfactual, the "control conditions," or what happens in the absence of the program. This evaluation logic is presented in Figure 2.2. The differences between what happens under the treatment condition and what happens in the absence of treatment (the counterfactual or control conditions) represents the "contrasts" that are the focus of evaluation inquiry. These contrasts exist in each part of the logic model. For example, the inputs (i.e., program design, implementation and administrative plans) to the program being evaluated differ from the inputs that exist under the control conditions.

There is also contrast between treatment and control conditions regarding the program activities: This box is often at the center of **implementation research**, which explains program operations, including the extent to which the program operates as planned (or, in other words, the program's **fidelity** to its design). This examination should take place both for the treatment logic as well as for the counterfactual, although the counterfactual logic is not commonly articulated or examined in practice. The assessment of "implementation fidelity" refers to the extent to which the theoretical program logic model is fully implemented in practice, and its counterpart in the counterfactual logic is "organizational context," or the operations of the world as it exists absent program efforts. To fully judge the vertical differences between the treatment and control logic models in Figure 2.2, the horizontal connections between logic model's elements must also be clearly assessed.

FIGURE 2.2 ■ Evaluation Logic Model

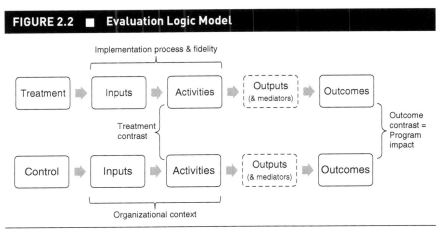

Source: Adapted from Weiss, Bloom and Brock (2013, p.14) with permission of MDRC. Copyright © 2013 by MDRC®. All rights reserved.

These horizontal connections also consider the intermediate pathways of interest that connect (or mediate) the program activities' to the target outcomes, the difference (or "contrast") between those target outcomes represents the program's impact. When a logic model more fully articulates the relationships between specific program activities and their outputs and hypothesized outcomes, those elements are positioned to be evaluated, to help answer questions about program improvement.

Weiss et al.'s (2013) justification for examining impact variation involves the observations that "(1) different types of programs (or different versions of the same program) vary in their effectiveness, (2) a program that is effective for one group of people might not be effective for other groups of people, and (3) a program that is effective in one set of circumstances may not be effective in other circumstances" (p. 1). Further, "policymakers and practitioners . . . want to know why some programs are more effective than others and what it might take to design and operate more successful programs" (Weiss et al., p. 2); and research to date has tended to be less able to answer these questions than questions of the average treatment effect.

To position an evaluation to help answer the "right" questions, I suggest using this framework to identify what it is about the program activities that is of specific interest. Once that is identified from a program's logic model, it can be juxtaposed with the control conditions as a way of identifying the "contrast" that is to be evaluated. For each of the program activities that are expected to produce certain program outputs and impacts, a clear and explicit logic model can identify the hypothesized relationships, and a corresponding counterfactual model can provide a means for understanding the relative impacts of those specific program activities.

CONCLUSION

The field of program evaluation is large, and there are many evaluation questions that hold value for program learning and program improvement. This book focuses on the questions that pertain not only to a program's impact but specifically to selected aspects of a program's impact. The general impact question is "What difference did the program make on improving the outcomes of interest?" Even when we just ask "What is the program's impact?" we need to be explicit regarding improving outcomes of interest "relative to what." The impact of a new program might be gauged relative to an old version of the program, or relative to an existing service environment that does not include the program, or relative to nothing (i.e., no services available at all). The evaluation logic model presented in this chapter provides a framework for being clear and explicit about contrasts of interest. The evaluation logic model presents the program logic model alongside counterfactual logic model, such that the differences between the two are the evaluation contrasts of interest.

In addition to the evaluation of wholesale programs, program logic models can and should be explicit about the component parts that might be the necessary ingredients to a program's success. The evaluation logic model provides a framework for articulating questions about how a program worked, for whom and under what circumstances. These additional questions might be posed as follows:

- What about the program is the reason for its observed impact (or lack of impact)? To what extent does the program's impact vary across subpopulations? To what extent does the program's impact vary across multiple sites? What components of the program are necessary for it to work? What other components are unnecessary for the program to work? What implementation strategies generate the greatest impact?

This chapter has presented a framework for positioning these questions. Adding the counterfactual logic to a program logic model creates an overarching evaluation logic model, which makes explicit the contrasts at the center of an **impact evaluation**. The conceptual framework lays out the dimensions of **implementation fidelity**, **treatment contrast**, and **outcome contrast** that evaluation can examine. This framework provides a basis for understanding which evaluation design options are suited to addressing each contrast.

QUESTIONS AND EXERCISES

1. Discuss: What is the difference between an "outcome" listed in a program logic model and an "impact" that might be estimated as part of an evaluation?

2. Create a logic model for a program you are familiar with.

   a. Supplement that logic model with a representation of the counterfactual logic (that is, the control conditions), and identify the areas of treatment and outcome contrast that might be of interest to an evaluation.

   b. Elaborate on the specific activities that are hypothesized to be the key ingredients of a program and demonstrate how the evaluation logic model can provide a framework for understanding the contributions of those activities to the program's impacts.

RESOURCES FOR ADDITIONAL LEARNING

W. K. Kellogg Foundation. (n.d.). *Logic model development guide.* Retrieved from https://www.wkkf.org/resource-directory/resource/2006/02/wk-kellogg-foundation-logic-model-development-guide

Weiss, M. J., Bloom, H., & Brock, T. (2013). *A conceptual framework for studying the sources of variation in program effects.* New York, NY: MDRC. Retrieved from http://www.mdrc.org/sites/default/files/a-conceptual_framework_for_studying_the_sources.pdf

3 THE BASIC EXPERIMENTAL DESIGN DEFINED

In experimental evaluations a lottery-like process randomly divides those eligible for treatment into two groups: a **treatment group** assigned to receive the program or policy that defines the intervention and a **control group** excluded from the program or policy for research purposes. The control group provides a powerful comparison group—or counterfactual—that tells us what would have happened in the absence of treatment.

Generally, in order to establish a cause-and-effect relationship, three criteria must be met (e.g., Babbie, 2016): The two events must be correlated, cause must precede effect in time, and no other explanation must exist. The first two of these are straightforward to judge, and it is the third that poses a research design challenge that impact evaluations aim to meet. A well-executed experimental evaluation allows us to rule out alternative explanations for why an intervention achieved its effects, eliminating historical forces, selection bias, maturation, regression artifacts, instrumentation bias, and interactions among these. These plausible rival explanations are called "threats to internal validity" and are defined as follows:[1]

- *Historical forces* (political, social, and economic). People find jobs more quickly in a strong economy than in a weak one. People feel more patriotic around election day. Specific violent events—such as school or mass public shootings—influence public acceptance of gun safety laws.

- *Selection bias.* People most likely to succeed in and benefit from a program are those who enroll. Selection can involve creaming of program participants by program staff, or it can involve participants' self-selecting to participate. The reverse of selection into a program is attrition out of a program.

- *Maturation.* People learn and grow over time, in ways that affect their outcomes. New programs mature over time into a steady administrative state, and that maturation process may be mistaken for program impact.

- *Regression artifacts.* People who enroll in programs to help them are there because they are at a low point in their lives and will get better ("regress"

---

[1] For a complete treatment of the concept of threats to an evaluation's internal validity—or the "reasons why we can be partly or completely wrong when we make an inference about . . . causation" (p. 39)—see Shadish, Cook, & Campbell (2002, pp. 53–61) or most any other evaluation text book.

or move toward the mean) even without a program's help. On the flip side, programs for children with high standardized test scores, for example, have little additional upward room to move those children, who are more likely to report worse outcomes by simple **regression toward the mean**.

- *Testing bias.* Being tested can sensitize a person and thereby influence their outcomes. For example, asking someone about their nutrition habits might make them more likely to make better nutrition choices, regardless whether they are subjected to some intervention aimed at influencing their nutrition.

- *Instrumentation bias.* Consistent measurement matters, and changing measures or data collection procedures over time can result in mistaking instrumentation bias for impact. For example, if community policing makes people more comfortable reporting crime, then a measure of criminal activity under an old versus new regime might make it look as if crime has increased, when it is simply reporting that has increased. The measure has become more sensitive or differently calibrated, when no real change has occurred.

- *Interactions* among these threats to internal validity exist as well: Selection of the "best" people to participate during an expanding economy brings selection, regression artifacts and historical threats combines to generate a complex interaction of factors, each of which—alone and together—might otherwise take credit for changes in outcomes, in lieu of true program impacts.

The only threat to internal validity that an experimental evaluation design does not deal with is that of differential attrition between a treatment and control group, sometimes referred to as experimental mortality. This occurs when the experimental evaluation design itself creates a treatment-control distortion in data collection, making one group more or less likely than the other to complete or be available for follow-up data collection.

BOX 3.1

IMPLICATIONS OF THREATS TO VALIDITY FOR A TUTORING PROGRAM EVALUATION

Consider a tutoring program for second-grade children who are reading at the first-grade level. Parents can enroll their children in tutoring in order to bring their reading skills up to grade level. A **pretest–posttest** evaluation design showed that the program increased reading at grade level 5,000%! At the beginning of the school year 1% of the children in the program were reading at grade level, and by the end of the year 50% were. Three main alternative explanations exist:

- *Selection bias.* The parents who chose to enroll their children in tutoring could be more invested in helping their children outside of the program as well, contributing to their children's improvement in reading.

- *Maturation.* Young children are like sponges. They learn how to read in part through their own maturation processes, aided by other stimuli in their environments.

- *Regression artifacts.* Children reading below grade level have just one way to go: up. Moving toward the mean is another reason their reading scores might improve.

Instead of asking parents to enroll extremely poorly performing children into an intervention, if the evaluation would have randomized below-grade-level students into a tutoring program, then those with access to the tutoring would have the same characteristics as those in a control group. In turn, examining later differences in the outcomes between those with and without access to the program would net out the influences of selection bias, maturation, and regression artifacts. As a result, the evaluation would be able to tell a causal story about the effects of tutoring rather than be unsure of what portion of that 5,000% effect might be attributable to the program.

Moving beyond the average impact of offering tutoring, if teachers and school administrators have the resources to offer tutoring to their lowest-performing students regardless, then—instead of needing to know whether tutoring is better than no tutoring—they might be more interested to know what kind of tutoring (e.g., one-on-one, peer-led, small-group, computer-assisted) and what level of intensity (e.g., 1 hour per day, 1 hour per week) makes tutoring most effective. Experimental evaluation designs can also be used to introduce systematic variation that permits parsing out the key elements of a program to determine what is essential to its effectiveness, what features might increase its effectiveness, and/or what features are not necessary and might be cut in order to reduce the program's cost.

RANDOM ASSIGNMENT EXPLAINED

What exactly does **randomization** involve? The process of randomly assigning units—be they individuals, classes, schools, organizations, cities, and so on—to gain access (or not) to an intervention involves something akin to a coin toss or a roll of the dice. As shown in Figure 3.1, when eligible units are identified, they are subjected to randomization (a coin toss), which determines whether they gain access to the program or not. An evaluation then follows these two groups. Comparing their outcomes provides an estimate of the program's "impact." Because the only systematic difference between the groups the intervention, the difference between the two groups' outcomes can be interpreted as the intervention's causal impact, which is unbiased by other influences.

FIGURE 3.1 ■ The Randomization Process Is Like a Coin Toss

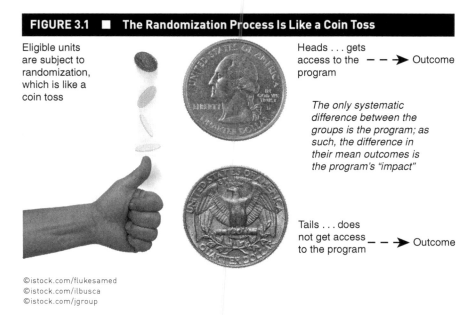

Eligible units are subject to randomization, which is like a coin toss

Heads . . . gets access to the program − − ➤ Outcome

The only systematic difference between the groups is the program; as such, the difference in their mean outcomes is the program's "impact"

Tails . . . does not get access to the program − − ➤ Outcome

©istock.com/flukesamed
©istock.com/ilbusca
©istock.com/jgroup

In practice, randomization does not actually involve a physical coin, although it could. Typically, a random assignment algorithm is built into a computer program that interfaces with the program's management information system so that when an individual—or a group of individuals or other unit of study—is deemed eligible, the algorithm decides which path that individual follows after the figurative or virtual coin has been tossed: the control group or the treatment group. The algorithm can be as simple as a spreadsheet-based random number generator, where resulting values over 0.50 gain an individual access to the program and resulting values of 0.50 and lower result in referral to the control (or alternative treatment) group.

This fifty-fifty assignment ratio results in treatment and control groups of equal size, but there are often reasons one might choose a different ratio. In some cases, budget constraints might limit treatment to less than half of the applicants, in which case the treatment group would be smaller than the control group. For example, a program's limited resources might mean that it can serve only a quarter of those who are eligible. In that case, three control group members would be designated for every one treatment group member. Conversely, sometimes program administrators want to minimize the numbers of individuals referred to a control group, and so having a larger treatment group might be desirable. This was the case in the national Job Corps evaluation. The treatment group included about 81,000 youth and the control group about 6,000: that is, the control group made up less than 7% of the research sample (Schochet, Burghardt, & McConnell, 2008). Having uneven treatment and control group sizes—where there can be either a smaller or larger treatment group—is common in practice. Further, as noted in Chapter 1, excess demand creates an ethical rationale for randomly selecting program participants.

A control group also can be offered an alternative treatment, in which case no one is excluded from services but instead research subjects receive access to alternative services. This kind of comparison lends itself to informing program improvement efforts, as elaborated later.

BOX 3.2
RANDOM ASSIGNMENT IS NOT RANDOM SAMPLING

Random assignment into an experimental group (treatment group and either a control or alternative treatment) should not be mistaken for random sampling. Random sampling pertains to the process by which a sample is drawn to represent some broader population of interest. In comparison, random assignment is the approach used to select from within a given sample units that gain access to a given treatment or not.

BOX 3.3
TWO ADDITIONAL FEATURES OF RANDOMIZED EXPERIMENTS IN PRACTICE

Blocking. Blocking involves forcing—over some preset number of randomizations—that the chosen ratio or treatment to control cases be met exactly. If one were to use a block of 20, and had a **random assignment ratio** of 1 to 1, then the largest number of consecutive control group (or treatment group) members that could arise would be 20: This would come about if the last 10 of one block were, for example, treatment cases, and the first 10 of the next block were treatment cases. Of course this is a very unlikely occurrence, but the block size can be set so as to prevent administrative frustration with long strings of one type of assignment or other. In an educational or job-training setting, where program administrators are trying to fill classes of a particular size by a particular date, it can be impractical to get many control cases in a row, when classes need to be filled in order to take place. In that case, a block size of fewer than 20 might be preferable. A block size that is too small, however, can result in program staff thinking that they can "game" the system, and "know" when a treatment of control slot is likely to be up next. As a result, it is best to keep details of the algorithm separate from the process of taking in study participants.

Stratification. In randomizing units within an experiment, groups called strata can be identified and used either to ensure population representation or to establish overrepresentation. There might be research or administrative reasons to give a greater (or lesser) chance to some kinds of people to be selected to treatment. For example, if public assistance recipients were a target population of interest but represented only a minority of program eligibles, then a random assignment algorithm might establish a greater probability for those individuals to be in the treatment group, thereby ensuring a large enough sample size for later subgroup analyses.

BOX 3.4
A SIDEBAR ON SEMANTICS

As Chapter 1 noted, the field of evaluation is large and diverse. Many types and styles of evaluation exist, a feature of the field that results in there being many labels for similar things. This Sidebar on Semantics identifies some alternative labels for experiments and offers an explanation for why I choose to use some and not others in this book.

- *Random Assignment Evaluation* or *Random Assignment Study.* These labels for an experimental evaluation are imprecise and incorrect. The process of random assignment is the means through which an evaluation creates treatment and control (or alternative treatment) groups. Random assignment is an essential feature of an experimental evaluation design, but it is not the design itself.

- *Randomized Control Trial* or *Randomized Controlled Trial* (RCT). The RCT label is commonly used in medical and pharmaceutical studies. In recent years, these labels have gained favor in policy evaluations. It is my hypothesis that this has happened for two reasons: desire to enhance the perceived rigor of experimental evaluation research in public policy and also to establish an acronym.

- *Randomized Experiment* or *Experimental Evaluation.* These two terms are my preferred labels for evaluations of policies and programs that use randomization to allocate access between treatment and control or among alternative treatment groups.

- *Classically-designed Evaluation* or *Classically-designed Experiment* are other appropriate labels for evaluations that involve randomization of units into experimental groups.

Another important semantic point pertains to the use of the term *control.*

- A *control group* is one that is randomized to and therefore equivalent in all ways to a randomized-to-treatment group.

- The term *experimental group* refers to a control or treatment group (or alternative treatment groups) that has been formed via a randomization process.

- A *comparison group* is created through nonrandomized processes and may also be referred to as a "non-equivalent control group" (e.g., Shadish, Cook, & Campbell, 2002).

Other, related kinds of evaluations fall under the following labels:

- *Natural Experiments.* A class of evaluations called *natural experiments* involves allocating access to or exclusion from a treatment group by natural forces, those not manipulated by a researcher but that emulate random assignment. In the United States, cross-state implementation of seat belt laws or legal drinking age, or lotteries operated to determine school enrollment are examples.

- *Efficacy* versus *Effectiveness Trials.* An efficacy trial is an evaluation of an intervention "under ideal and controlled circumstances" whereas an effectiveness trial is an evaluation of an intervention under "real-world" conditions (e.g., Singal, Higgins, & Waljee, 2014).

THE BASIC (TWO-ARMED) EXPERIMENTAL DESIGN

Three slight variants of a basic, two-armed experimental evaluation design can be summarized as follows:

- Treatment vs. Control, where control involves no services

- Treatment vs. Control, where control is a **status quo**, **business as usual** or **existing service environment**

- Treatment vs. Alternative Treatment

Treatment vs. Control (no services)

The first of these—the treatment group versus a **no-services control group**—involves randomization of units to (1) gain access to a treatment or (2) be strictly excluded from the program. This "no services" control group represents a counterfactual where *no alternative services* are available, as might be the case in a new kind of program for which there are not existing alternatives. Examples of this come from experiments in the developing world, where people or villages are randomized to gain access to a new benefit and nothing similar is available outside of the experiment. To test whether no-strings-attached cash payouts could indeed lift people from poverty, an experiment in Zambia gave families $18 per month bimonthly for 2 years (Aizenman, 2017a; Handa, Natali, Seidenfeld, Tembo, & Davis, 2018) and another experiment in Kenya randomized villages—and all families within them—to receive $22 per month, to continue for 12 years (Aizenman, 2017b). These potentially life-changing amounts went only to those randomized into treatment, with nothing being allocated to a control group. This is because "nothing" really does represent the counterfactual state. Using this evaluation design, researchers can learn about the impacts of making these payouts without worrying that other factors are influencing results. In contrast, in the United States, very rarely are educational and social policy control groups excluded from access to *any* services. Instead they are simply excluded from participating in the program being evaluated.

Treatment vs. Control (Status Quo, "business as usual" or Existing Service Environment)

The treatment group versus a *status quo* ("business as usual" or existing service environment) control group is a common application of the basic two-armed experimental evaluation design. In this situation, eligibles are randomized to gain access to a program, but those randomized to the control group still have access to any services that are available in the community. This is the case with Head Start, for example, where early childhood education and child care programs of a wide variety are available throughout the community. As such, an evaluation of Head Start—as was implemented in the Head Start Impact Study in the early 2000s (Puma et al., 2012)—considered the extent to which gaining access to Head Start improves children's kindergarten readiness, relative to the constellation of alternative options that exists. Similarly, current job training programs do not assess the effectiveness of training compared to nothing, but instead they assess the effectiveness of a specific model of training compared to whatever other alteratives are out there. Of course, some communities will have greater training resources than others, and understanding what specific services represent the counterfactual is an important part of any evaluation. In the case of entitlement programs, when people cannot legally be denied access, evaluations often consider policy alternatives to the status quo policy or program. This status quo is commonly referred to as representing "business as usual" or "usual care."

Treatment vs. Alternative Treatment

There are cases in which neither of these first two design variants—those with a no-services or a **status quo control group**—answers the question of interest. The third variant of the basic two-armed experiment is what the business world refers to as "A/B" testing (e.g., Manzi, 2012; Thomke & Manzi, 2014). This involves randomizing to two kinds of treatment, such that neither group is strictly a control (either no services or status quo). This A/B testing involves comparing alternative treatments to ascertain which works better, relative to one another.

This third variant of the basic sometimes involves comparing alternative treatment models, adding a specific enhancement to a program, and scaling up or back a program. A "competing treatments" design would compare fully alternative treatment models. An "enhanced treatment" design would allow a program to test whether the enriched—or scaled back—services meaningfully changes the program's impact before committing to making the change. The scale-up or scaleback option permits assessing what levels of services or assistance result in what levels of benefits.

These Treatment vs. Alternative Treatment designs can be used in conjunction with ongoing program operations, allowing managers and administrators to learn about the effects of potential changes in a deliberate way. Common in the business world as A/B testing (e.g., Brooks, 2012; Manzi, 2012), Alternative Treatment vs. Treatment evaluations are not (yet) common practice in the government or nonprofit sectors. That said, it is my hope that this book's lessons will contribute to changing that by highlighting the important opportunities for program and policy learning that come from this and other designs.

TO HAVE A CONTROL GROUP OR NOT TO HAVE A CONTROL GROUP?

The two-armed Treatment vs. Alternative Treatment design involves the comparing of outcomes between two treated groups. Must one of these groups represent the status quo (or "business as usual" or "usual care")? Is an excluded control group really necessary? On the one hand, the control benchmark is needed to determine whether any of the alternatives tested improve on current policy. However, if a funder is committed to running an intervention (no matter the research results), then testing multiple interventions against an excluded control group is not necessary. Instead, the evaluation can focus on comparison of alternative configurations of the intervention or vary certain aspects of the program in an effort to detect the most essential or impactful ones.

A recent example of this is the Job Search Assistance (JSA) Strategies evaluation, mounted by the U.S. Department of Health and Human Services (HHS; n.d.b). In this case, HHS was interested in learning *what about* JSA is most effective for moving welfare recipients to work. It is not a viable policy option to provide *no* JSA—those receiving welfare must engage in some activities aimed at finding work—and so the evaluation challenge is to discern which JSA strategies appear to be most effective in helping recipients find work in general, find work more quickly, or find better work.

Another example is that of the Family Options Study that considered alternative modes of providing services to homeless families (U.S. Department of Housing and Urban Development [HUD], n.d.). This is a situation where randomizing to a no services control group was deemed unethical, That is, it is widely accepted that society *cannot* permit children to be homeless and therefore *must do something* for homeless families who ask for help. In response, the evaluation randomized families to four alternative treatments in order to ascertain which is most effective in improving the outcomes of this vulnerable group. Each of the treatments provided at least what was then available to homeless families. In addition to a usual care group (which involved access to existing housing and services), the evaluation randomized families to a subsidy group (who received a permanent housing subsidy in the form of a voucher), to a project-based transitional housing group (who received temporary housing for up to 24 months with an intensive package of supportive services), or to a community-based rapid re-housing group (who received temporary, potentially renewable rental assistance, along with some housing-focused services). Without a "no services" control group, we cannot say whether any of these options is better than nothing; but nothing is not a socially acceptable alternative to test. Instead, the study has been important in identifying how best to help homeless families. Without a control group, the study provided

> striking evidence of the power of offering a permanent subsidy to a homeless family. Families who were offered a housing voucher experienced significant reductions in subsequent homelessness, mobility, child separations, adult psychological distress, experiences of intimate partner violence, school mobility among children, and food insecurity at 18 months. Moreover, the benefits of the voucher intervention were achieved at a comparable cost to rapid re-housing and emergency shelter and at a lower cost than transitional housing. (Gubits et al., 2014, p. iv)

The model of randomizing to alternative program configurations might be especially appealing to nonprofits that have program support yet want to engage in program improvement efforts.

It is my contention that the control group as the field has come to think of it—that is, a no services or **business as usual control group**—is largely responsible for experiments' negative reputation among program managers. It does not need to be the common model for experiments in practice. Program evaluation has evolved sufficiently that only very rarely are the important questions about a program's results as compared to nothing. Instead, policymakers and practitioners alike are much more interested in programs' essential ingredients. The designs that I present in Chapter 4—while they can involve a conventional control group—are amenable to answering questions about those essential ingredients, that is about what aspects of a program drive impacts.

With this fundamental understanding of a two-armed experiment, Chapter 4 explores additional design options—such as additional arms, time-staggering, and combining program components in varied ways. These design variants provide options for examining the relationship between various program configurations and impacts.

QUESTIONS AND EXERCISES

1. What are the three variants of a basic, two-armed experimental evaluation design?

2. Provide an applied example of each variant or propose where you might implement each design in practice.

3. Discuss: What are some circumstances where an excluded control group (no services or business as usual) is unnecessary?

RESOURCES FOR ADDITIONAL LEARNING

Gerber, A. S., & Green, D. P. (2012). *Field experiments: Design, analysis, and interpretation*. New York, NY: Norton.

Manzi, J. (2012). *Uncontrolled: The surprising payoff of trial-and-error for business, politics, and society*. New York, NY: Basic Books.

Rossi, P. H., Lipsey, M. W., & Henry, G. T. (2004). *Evaluation: A systematic approach* (8th ed.). Thousand Oaks, CA: Sage.

Shadish, W. R., Cook, T. D., & Campbell, D. T. (2002). *Experimental and quasi-experimental designs for generalized causal inference*. New York, NY: Wadsworth.

4

VARIANTS OF THE EXPERIMENTAL DESIGN

This chapter presents several variants of the experimental evaluation design, variants that are suited to providing information about program design and improvement decisions. These designs include randomizing to three and four (and potentially more) groups, where those groups represent distinct alternative treatments of some sort. They also include alternative timing: rolling out an intervention across multiple locations over time, for example, so that all can gain access to the intervention, but the timing of when they do is randomly decided such that some gain access immediately while others serve as a control group before they gain access. After defining and offering examples for each design, I suggest how certain program characteristics lend themselves to certain designs.

These varied designs fall into the following categories:

- Multi-armed

- Factorial

- Multistage

- Staggered introduction

- Blended designs

MULTI-ARMED DESIGNS

Definition. As the name implies, a **multi-armed experimental evaluation** design involves randomizing to three or more groups, where those groups can include a control group and do include multiple variants of a treatment in order to ascertain which is preferable. Multi-armed experiments can compare distinct program models to one another (the "competing treatments" design), or they can compare alternative versions of the same program model (the "enhanced treatment" design). The latter of these—the enhanced treatment design—can consider either an enhancement or scale up to a given program (as described in the examples that follow), but it can also involve scaling back as an option. A scale up option implies adding something to a program and testing whether the addition confers benefits; whereas a scale-back option implies taking something away from a program and testing whether the subtraction reduces the program's effectiveness. Either way, the treatment arms compare a standard and one or more modified versions of the same program.

Competing Treatments Design and Examples. The National Evaluation of Welfare to Work Strategies (NEWWS), which operated in the 1990s, is an example of the "competing treatments" design. It was a multi-armed experiment that compared two alternative program models (Hamilton et al., 2001). The evaluation randomized applicants who were eligible for welfare assistance to a control group or to one of two distinctive treatment models. In one treatment, individuals had access to welfare assistance based on a labor force attachment (LFA) model. This model emphasized "work first," which encouraged participants to find any job on the premise that building labor market experience would help them into better jobs later. In the other treatment arm, individuals had access to welfare assistance based on a human capital development (HCD) model, which emphasized education and training as the pathway to better labor market outcomes. Each of the several local welfare offices that implemented the study operated both program variants plus a control group that did not receive any welfare-to-work assistance. This design allowed the study to answer questions about the extent to which either program model improved on a welfare program's business as usual without work requirements (LFA versus control, HCD versus control) as well as the extent to which the two approaches' impacts differed from one another (LFA versus HCD).

A head-to-head test of program models of this sort is relevant to program administrators in that it informs their choice of which model to operate; and it is relevant to policymakers and funders because it informs their choice of what to fund.

The NEWWS compared two alternative program models against each other and against a status quo control group, but—as noted in Chapter 1—a control group is not strictly necessary. The Family Options Study described in Chapter 3 is another example of a competing treatments design. It compared standard treatment of homeless families with three alternative models to ascertain whether we, as a society, should be changing the way we treat this vulnerable population. That evaluation concluded that we should (Gubits et al., 2018).

Enhanced Treatment Design and Examples. In comparison to the "competing treatments" design, the "enhanced treatment" design involves either scaling up or scaling back a program in order to compare the base version of an intervention to an alternative (either enhanced or reduced) version of that program. This approach measures the contribution of the added component, for example, to the intervention's impact magnitude. The Social Security Administration's (SSA) ongoing Benefit Offset National Demonstration (BOND) illustrates this approach. The study contrasts a more generous way of paying Social Security Disability Insurance (SSDI) benefits in one experimental arm to that same benefit adjustment plus enhanced employment counseling in a second experimental arm. The study also compares both treatment arms to a business as usual control group, which represents standard SSDI payment rules (Gubits et al., 2014). This design allows policymakers to learn whether the benefit payment change affects individuals' earnings compared to the current rules, whether the payment change plus enhanced counseling affects earnings compared to the current rules, and whether the enhanced counseling itself increases the impact of the new benefit rules.

Another way to develop an enhanced treatment design—in the context of a many-location program—is to give each of a set of local program agencies the autonomy to choose an enhancement to its base program that can be accessed via a lottery. With this design, a single experiment that operates in several locations can test multiple enhancements, as is the case with the Health Profession Opportunity Grants (HPOG) program's impact evaluation (Peck et al., 2014, 2018a). The HPOG program offers career-pathways-based training for health care sector careers; and, the evaluation of its first round of funding considered 42 distinct programs' operations in more than 90 locations. The evaluation carved out three specific program enhancements that were added to the base program in some of those locations for an experimental test. Three grantees enhanced their base programs by offering facilitated peer support groups, intended to increase program attachment and improve retention. Three other grantees tested the effect of offering additional financial assistance to people who experience exceptional financial stress, such as the threat of imminent eviction from their homes. An additional five grantees tested a system for rewarding participants for achieving particular program milestones such as perfect attendance, high grades, or staying in a job for at least 90 days. By randomizing individuals to the base program and to one of these program enhancements, the evaluation could assess whether peer support, emergency assistance, or non-cash incentives provide—across its many sites—a meaningful boost to the HPOG program's overall effectiveness.

Although the HPOG enhancements test is part of a national evaluation, the programs that chose to offer one of these enhancements via a lottery were local. As such, they offer an example of how single, local programs could use the enhanced treatment design to inform their own program management decisions. Programs are rarely, if ever, static. Indeed, program administrators are continuously considering ways to improve, streamline, or expand their services. As they consider these changes, the enhanced treatment evaluation design should be a go-to in their toolkit. By taking the time to evaluate—through a rigorous design that will ensure high quality evidence—administrators will have the information they need to feel confident in the decisions made regarding whether to extend a given enhancement programwide.

Next, consider an example of an enhanced treatment design, where one of the treatment arms scales back on an aspect of the program, in order to judge whether the same outcomes can be achieved but at a lower cost. In the first two summers (2011 and 2012) of the Summer Electronic Benefits Transfer for Children (SEBTC) Demonstration, the USDA tested providing $60 per month to families who would otherwise have gotten no food assistance during the summer. Then, to learn about the sensitivity of the benefit amount to the program's impacts on food insecurity outcomes, in 2013, it tested a scaled back benefit of $30 per month. This test revealed no difference in impacts for the lower benefit amount, implying that the USDA could spend less and achieve the same results (Collins et al., 2015).

Multifaceted programs interested to learn whether certain aspects of their programs are essential can devise plans for carving out and scaling back on the delivery of selected aspects, following a treatment group's outcome to ascertain whether "less is more," whether any program efficiencies imply more favorable outcomes, or whether the same results can be achieved with fewer resources. I recognize that program administrators may be hesitant to scale back on their offerings, but I raise this variant of the enhanced treatments design as a viable option for being deliberate in figuring out how to respond to possible funding shortages. Under threat or reality of funding cuts, programs could take the opportunity to experiment and learn how not to compromise impacts, if possible, while streamlining service delivery in some ways.

FACTORIAL DESIGNS

Definition. In a **factorial design**, selected aspects of a treatment or "factors" as they are labeled in this design, each of which has two (or more) levels, are randomized separately. Levels can be low or high dosage or simply absence or presence of the factor. As Collins et al. (2005) elaborate,

> which program components are working well; which should be discarded, revised, or replaced; which dosages of program components are most appropriate; whether delivery components are enhancing, maintaining, or diluting intervention efficacy; and whether individual and group characteristics interact with program or delivery components. (pp. 65–66)

are important questions—usually more important than the average treatment effect—that a factorial design can aid in answering. When the research question asks what it is about a program that is working, or not, or could be refined to be more effective and efficient, this design approach seems especially appropriate.

In its simplest form, the factorial design varies two treatment dimensions or factors, randomizing to each individually and to both together. If the levels of each factor include "absence" or "presence," then the absence of both factors represents a status quo control group. These factors are program components that can be carved out from the program as a focus of evaluation. A factorial design could accurately be considered a four-armed experiment, where two alternative service models (identified in Figure 4.1 as Cell 2 and Cell 3) offer access to one of two distinct alternative intervention options and the third alternative service model (Cell 4) offers access to both of the alternative options.

This factorial design offers an opportunity to detect the overall impact of each factor by using a "control" group for a given factor that includes all of those

FIGURE 4.1 ■ Illustration of 2X2 Factorial Design

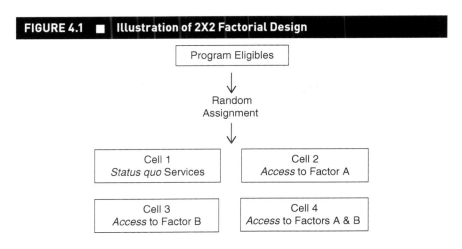

randomized *not* to have access to that factor. For example, referring to Figure 4.1, there are two ways to estimate the effect of Factor A: We can compare the combination of Cell 2 and Cell 4 (where Factor A is present) to the combination of Cell 1 and Cell 3 (where Factor A is absent); or we can compare the outcomes of those in two single cells, Cell 1 (where only Factor A is present) and Cell 2 (where neither Factor is present). The first of these maximizes the sample and estimates the impact of Factor A where individual experiences are a blend of the presence and absence of other factors. The second of these involves a comparison of Factor A to the control group only: This is a "purer" comparison but it also involves a relatively smaller sample size. As Shadish, Cook, and Campbell (2002) describe a factorial design (p. 264), each participant has the potential to do "double duty" by the nature of his or her exposure to particular factors or factor intensity levels. As such, some comparisons can involve relatively larger samples, which permit the design to answer additional questions, depending on which cells are combined to create the contrast of interest, as relevant to the research question.

Factorial designs are configured to support estimation of the incremental (marginal) impacts of individual factors or their intensities and to estimate the combined impact of the two factors or their intensities—an analysis that can reveal whether the combined factors are synergistic or, together, less than the sum of their parts. Although the Social Security Administration's BOND evaluation described earlier used a three-armed design, it very well could have used a factorial design. In doing so, it could assess separately the effect of adjusting the benefit amount, the effect of counseling services, and the effect of combining the new benefit amount with counseling.

A 2x2 factorial design involves four cells but can support answering eight questions, as follows, using the full sample (for Questions 1 and 2) or single cell/pair comparisons (for Questions 3 through 8):

(1) What is the effect of Factor A? (compared to non-A)

(2) What is the effect of Factor B? (compared to non-B)

(3) What is the effect of adding Factor A on top of B?

(4) What is the effect of adding Factor B on top of A?

(5) What is the effect of Factor A? (alone, compared to C)

(6) What is the effect of Factor B? (alone, compared to C)

(7) Which of Factor A and Factor B is more effective?

(8) What is the effect of Factors A & B together? (compared to C)

Next, the design represented in Figure 4.2 depicts the use of intensity in the factorial matrix. As such, this design can inform the extent to which Factor A at high intensity, Factor B at high intensity, or the combination of Factors A and B at high intensity is more effective than having both factors at low intensity. A 3x3 design involves nine cells and supports answering 16 questions. A **fractional factorial design** involves testing a subset of all of the possible hypotheses available from the full matrix, by not randomizing to one or more of the possible cells.

Randomizing into a factorial design is not technically complex, but it may pose administrative or implementation challenges. The most likely obstacles to using a factorial evaluation design in practice is the host organization's capacity to run separate program variants and its willingness to do so. That is, agencies offering these alternative services must be able to ensure that the individuals randomized to each cell actually receive the services associated with that cell and that there is no crossover

FIGURE 4.2 ■ Illustration of Factorial Design with Varying Intensity

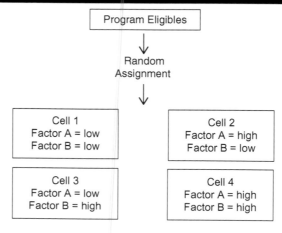

between and among them. **Crossover** refers to people being randomized into one group actually gaining access to the services provided to a different group. Minimizing crossovers ensures the design's ability to answer the research questions about the independent and interaction effects of the two factors. Although relatively uncommon at this time within the social welfare policy arena, factorial designs are more commonly used in the health care and health policy arenas. For example, the work of Collins et al. (e.g., 2005) highlights how health interventions use factorial evaluation designs to parse out the extent to which medication and counseling contribute independently and together to improve health outcomes.

Examples. A well-known application of this design is the New Jersey Negative Income Tax experiment in the late 1960s and early 1970s, where income guarantees were defined by poverty level (at 50, 75, 100, and 125%) and the tax rate on earnings (30, 50, and 70%). Although this 4x3 design would have 12 possible cells (and the ability to answer many more questions because of the various interaction effects), researchers assigned participants to only the eight least expensive and most politically feasible combinations (Kershaw & Fair, 1976).

Medical trials and evaluations in the health sector more commonly use factorial designs in practice than is the case in the educational and social program arena. That said, assessments of the effectiveness of communications strategies offer a desirable setting for factorial designs. For example, a recent U.S. Occupational Health and Safety Administration (OSHA) study considered the relative and combined effectiveness of e-mail and mail communications with worksites. The On Site Consultation Marketing Study tested the effect of various program brochures (and no materials), where each brochure was tested without and with e-mail follow-up. The study revealed that *any* communication increased workplaces' requests for training, and the *specific type* of communication was irrelevant, neither contributing to nor suppressing impacts in any detectable way (Juras, Minzner, & Klerman, 2018).

Factorial designs are not yet as widely used in educational and social policy evaluations as they might be; but increasing demand from funders to address questions regarding *what about* programs drives their impact implies opportunities for the future. Considering a multifaceted program such as HPOG, a factorial experimental evaluation design could capture either the presence or absence of two (or more) program enhancements or the intensity of two (or more) components of the intervention, in both cases either with or without a control group. Two of the enhancements that were part of the HPOG Impact Study's three-armed experimental design could be brought together in a factorial design to consider their synergistic effects. For example, the emergency assistance enhancement offers individuals up to $1,000 to help them respond to emergencies, such as imminent eviction or necessary car repair. But what if the evaluation would have designed the evaluation to test levels of emergency assistance, and the extent to which such assistance interacts with participating in facilitated peer support? A factorial design that randomizes to one binary factor (e.g., yes/no access to peer support) and one intensity-defined factor (e.g., access to no, low, medium, or high amounts of emergency assistance) would result in eight cells and permit analyzing the effect of offering peer support and its interaction with the level of emergency assistance available to treatment group members, above and beyond what the base HPOG program offers.

MULTISTAGE DESIGNS

Definition. In a **multistage experimental design**, units placed in a particular group in an initial random assignment are then randomly assigned a second time into a second-stage treatment after they reach a certain trigger point. This design randomly assigns individuals to treatment or control at intake, subsequently observes eligibility for an aspect of the intervention in the treatment group, and then randomly assigns eligible individuals from that arm to gain access to the second stage intervention or not.

One particular type of **staged evaluation design** involves adaptive treatments and has been coined a "**sequential, multiple assignment, randomized trial**" (SMART; e.g., Murphy, 2005). This occurs when a participant's initial responsiveness to treatment determines later treatment options, considered adaptive treatments. More generally, a staged design uses an individual's initial treatment experiences or outputs—or even simply the passage of time—to determine later treatment options, to which individuals are randomized. Adaptive designs in particular can help assess the optimal sequencing of treatments or aspects of treatment as well as frequency or type of contact. For example, in a job training program, those randomized into treatment who do not find a job on their own within the first 6 months might be randomized a second time into more intensive job search services. In the case of more specialized job training programs, such as those operating in a career pathways framework, this is relevant because of the high level of customization that takes place: Programs often aim to individualize services. This design provides the possibility of randomizing participants to alternative and responsive treatment options across the lifespan of a program. As such, it permits ongoing learning that programs can incorporate into their decisions about design and implementation.

Figure 4.3 depicts how staged randomization works. By comparing the outcomes for those in the bottom left box (the Stage 2 "treatment" group) to the outcomes of those in the bottom right box (those in the Stage 2 "control" group), this design permits identifying the causal effect of having access to the Stage 2 opportunity among those who hit the eligibility trigger point. The ineligible (i.e., non-triggered) treatment group members are not included in this second-stage impact calculation.

In addition, the design also permits estimating the effect of the intervention overall, absent the second-stage opportunity. This evaluation contrast would compare the outcomes of the treatment group sample in the boxes along the left versus the first-stage control group outcomes.

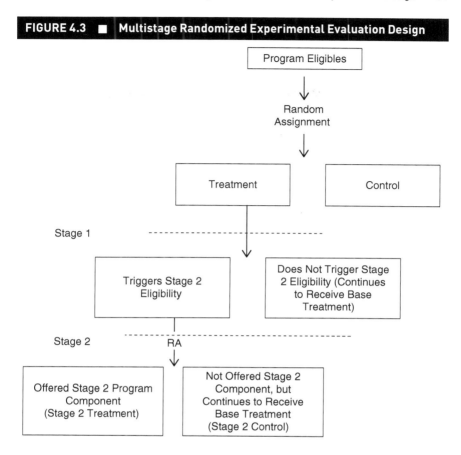

FIGURE 4.3 ■ Multistage Randomized Experimental Evaluation Design

Examples. Considering a program such as HPOG again, the staged design would be relevant for assessing the impact of a program component such as internship or on-the-job training (OJT) opportunities. These are aspects of the program both that have limited availability and where program administrators would likely want to impose an additional screen on who gains access. For example, only participants who complete some training and demonstrate readiness for the work environment might pass the threshold that would gain them access to one of these added opportunities: internships or OJT. By defining eligibility explicitly and then randomly assigning in a second stage, the design creates a group of individuals who were eligible to be part of that program component and did not receive it. In this circumstance—with additional eligibility criteria and limited availability—a lottery makes sense and can help answer the question about whether engaging in the second stage opportunity (e.g., internships or OJT) increases the effectiveness of training overall.

Another example of an adaptive design involves students' training attendance (Peck et al., 2018b). Particularly among more disadvantaged populations, simply attending training can be challenging. An adaptive intervention and associated evaluation could follow various strategies. For instance, the treatment could involve a counselor checking in with each trainee on a regular basis, where *regular* is defined by the student's response to treatment. An adaptive experiment to improve attendance and guard against dropping out might involve randomizing trainees to receive contacts once every 2 weeks from a counselor (the treatment group) or not at all (the control group). After 2 months, attendance rates could be reviewed. Treatment group members whose attendance was deemed insufficient (e.g., were absent more than once a week) would be re-randomized into two groups: One could receive contact every week from a counselor and the other would receive contact every 2 weeks (the control). Those whose attendance was deemed sufficient also could be re-randomized into two groups: one to get contact every 2 weeks, and one for which contact was scaled back to once a month. After an additional 2 months, the impact of these adaptive interventions could be measured. This design permits testing the ideal frequency of contact and the sequencing of certain program experiences. To be administratively feasible, the implementation of an adaptive design could be automated, preventing staff from having to figure out who gets contacted when and ensuring staff adherence to the experiment's protocols: Simple actions (set up to respond to the randomization results) can be set up to trigger responses (both frequency and content) on staff calendars.

STAGGERED INTRODUCTION DESIGNS

Definition. Most generally, **staggered introduction evaluation designs** can be used when an intervention can be rolled out across multiple locations or cohorts over time, with later starters serving as the control group for the first starters. This is a design that involves staggered, sequential rollout of an intervention, essentially creating a "waiting list" at random, giving all participants, clustered into groups, access over time. This can occur in school settings where a district might want to learn about the implementation procedures as well as treatment impacts of a program before deciding whether to expand it to all schools. A randomly selected subset of schools can serve as early implementers, to inform later implementation.

Staggered designs are ideal for testing modifications to programs' designs or administrative procedures, using early results to modify future iterations. The staggered introduction means that new program components can be tested among a subset of locations or a subset of participants to provide insights regarding whether full commitment to the new program configuration is warranted. Similarly, as in the case of the nudge or opportunistic experiments described in Chapter 1, tweaks to programs' administrative processes can be examined incrementally and refined, using staggered introduction as a means to identify whether further expansion of the change is desirable.

TABLE 4.1 ■ Staggered Introduction Experimental Evaluation Design					
Time	**Group 1**	**Group 2**	**Group 3**	**Group 4**	**Group 5**
Year 1	Treatment	Control	Control	Control	Control
Year 2	Treatment	Treatment	Control	Control	Control
Year 3	Treatment	Treatment	Treatment	Control	Control
Year 4	Treatment	Treatment	Treatment	Treatment	Control
Year 5	Treatment	Treatment	Treatment	Treatment	Treatment

Taking the staggered introduction design further is the **stepped-wedge design**, so named for the visualization of randomization across multiple groups (see Table 4.1).This staggered access to the intervention randomized across multiple groups takes place so that, eventually, the entire pool of eligible participants gains access (e.g., Hussey & Hughes, 2007). The timing of access is determined randomly, and, except for the first group of treatment cases, most in the sample serve as control cases for some window of time. The design can be used with groups of individuals or other naturally occurring clusters, such as classrooms, clinics, schools, or communities. Because all units eventually gain access to the intervention, this design has the potential to face less political or administrative opposition. Including a transition period in between "steps" permits implementation lessons to be processed and conveyed and impact analyses to take place to inform ongoing support for the intervention's expansion.

Another option is to offer different treatments to each group in the stepped-wedge design to further assess program improvement efforts. Doing so allows making cross-treatment comparisons so that the more effective treatment options can be implemented thereafter.

Examples. For an example of how this design might be used in practice, consider the case of American Job Centers' (AJC) case workers and their interactions with their clients. A common first step for AJC clients it to undergo an assessment of their employability, job aptitude, and job readiness (as in Peck & Scott, 2005). If an agency wanted to determine how best to improve its assessment tool and to evaluate the tool's effectiveness before a broad rollout, it could randomly assign certain offices or case workers to try a new assessment tool. There are costs associated with changing application procedures such as these, including staff training required to collect and use data from such assessment tools, so a staggered introduction approach makes administrative sense. Once a "better" tool is identified, it could be implemented systemwide or take advantage of the ongoing staggered implementation for efficiency and policy learning purposes.

Hussey and Hughes (2007) report on an example of Washington State implementing partner notification intervention to expedite treatment of selected sexually transmitted infections (STIs) across all of its 24 counties. A small individual experiment suggested the intervention's effectiveness, which justified the statewide test. Four groups of six counties, each separated by 6 months, were randomized to implement the intervention and collect and analyze data on prevalence of selected STIs. The evaluation ultimately provided evidence of the population-level effectiveness of the partner notification intervention.

BLENDED DESIGNS

Definition. The final category of designs is what I will call "blended designs." Blended designs are those that take design features from more than one of the designs presented above. For example, possible blended designs could combine a multi-armed design, where one of the arms also includes a second stage, or could combine the factorial and staged or staggered introduction designs. At intake, for example, the latter design (blended factorial and staged) would randomly assign individuals to treatment and control groups as defined in a factorial matrix, where one or more of those cells might also offer a second stage treatment opportunity. As Shadish et al. (2002) encourage adding "design features" to improve the causal inference from a variety of quasi-experimental evaluation designs, I encourage adding experimental design features to increase the number and type of questions that experiments can answer. The specific blended design that might be used will depend on the specific research questions being asked.

ALIGNING EVALUATION DESIGN OPTIONS WITH PROGRAM CHARACTERISTICS AND RESEARCH QUESTIONS

With these many design options available, how might program administrators and program evaluators identify which evaluation designs are best for answering which evaluation questions of interest? Table 4.2 summarizes the kinds of designs and evaluation contrasts with the evaluation questions that they are capable of answering. As noted, these experimental evaluation designs can have an excluded control group or not. Having a control group leads to questions about the overall effectiveness of a whole intervention; whereas having multiple treatment groups lends itself to examining questions related to program improvement. Rather than asking *does it work*, the competing treatments and enhanced treatment designs, for example, carve out aspects of treatments (in whole or in part) to focus on and thereby associate with questions about *what works*.

What are the program (or program component) characteristics that determine which evaluation designs are most relevant for teasing out the relative contribution

(*Text continues on p. 47*)

TABLE 4.2 ■ Impact Evaluation Questions and Relevant Design Options

Design	Excluded Control Group?	Contrasts	Impact Evaluation Research Questions
Two-armed	Yes	Program v. Control	• What is the effect of the Program (relative to no services)?
		Program v. Existing Service Environment	• What is the effect of the Program (relative to whatever else exists out there)?
	No	*Two-armed "Competing Treatments" Design:* Model A v. Model B	• What is the effect of Model A relative to Model B? (This could be New Program relative to Old Program.)
		Two-armed "Enhanced Treatment" Design: Base Program v. Program with Added Component	• What is the effect of adding the Component to the Base Program?
Three-armed	Yes	*Three-armed "Competing Treatments" Design* (where control group is no services or existing environment): Model A v. Model B v. Control	• What is the effect of Model A? • What is the effect of Model B? • Which of Model A and Model B is more effective?
	No	*Three-armed "Enhanced Treatment" Design* (where control group is no services or existing environment): Base Program v. Enriched Program v. Control	• What is the effect of the Base Program? • What is the effect of the Enriched Program? • What is the effect of the Enrichment (relative to the base program)?
		Three-armed "Enhanced Treatment" Design (where control group is existing program): Base Program v. Program with Component A v. Program with Component B	• What is the effect of Component A? • What is the effect of Component B? • Which of Component A and Component B is more effective?

(Continued)

TABLE 4.2 ■ (Continued)

Factorial	Yes	*Binary 2x2 Factorial Design:* Cell 1 = Control Cell 2 = Factor A Cell 3 = Factor B Cell 4 = Factors A & B *Can extend to larger matrices and consider fractional options*	Using the full sample: • What is the effect of Factor A? (compared to non-A) • What is the effect of Factor B? (compared to non-B) Using single cell pairs (reduced power): • What is the effect of adding Factor A on top of B? • What is the effect of adding Factor B on top of A? • What is the effect of Factor A? (alone, compared to C) • What is the effect of Factor B? (alone, compared to C) • Which of Factor A and Factor B is more effective? • What is the effect of Factors A & B together? (compared to C)
	No	*Intensity-based Factorial Design* Cell 1 = Base Program Cell 2 = Base + Factor A Cell 3 = Base + Factor B Cell 4 = Base + Factors A & B	Same as above as extended to consider levels of intensity

Staged	No*	Stage 1: Program v. Control *Trigger* (for Program group) Stage 2: Program v. Program + Option	• What is the effect of the Program? • What is the effect of the Option? • What is the effect of the Program with the Option?
		Stage 1: Program v. Control *Treatment Response* (for Program group) Stage 2: Alternative treatment contingent on response	• What is the effect of the Program? • What is the effect of following alternative treatment Option 2 v. staying on course/Option 1? • What is the effect of the Program including staying on the course? • What is the effect of the Program including changing course?
Staggered Introduction (Stepped Wedge)	No*	Program v. Control or Program v. Enhanced Program	• What is the effect of the Program? • What is the effect of the Program Enhancement?
Blended	Yes/No	Contrasts dictated by question(s)	Questions as relevant to program information needs

Notes:

#Although this is a design option, in application comparing alternative treatment models without a control group is not recommended unless the program's impact (relative to a no services or existing environment counterfactual) is not in question. Neither model might be effective (relative to no services or to the existing service environment), which then makes irrelevant their effectiveness relative to each other. Where this is a New Program compared to the Old Program, a two-armed competing treatments design is more relevant.

* In the staged design, a "control" group may be part of the first randomization, but the contrast of interest related to the second (staged) randomization considers a treated group (those randomized into the first-round services). In the staggered introduction designs, the "control" group will ultimately gain access to treatment.

FIGURE 4.4 ■ Relationship between Program Characteristics and Experimental Evaluation Designs: Applied Example

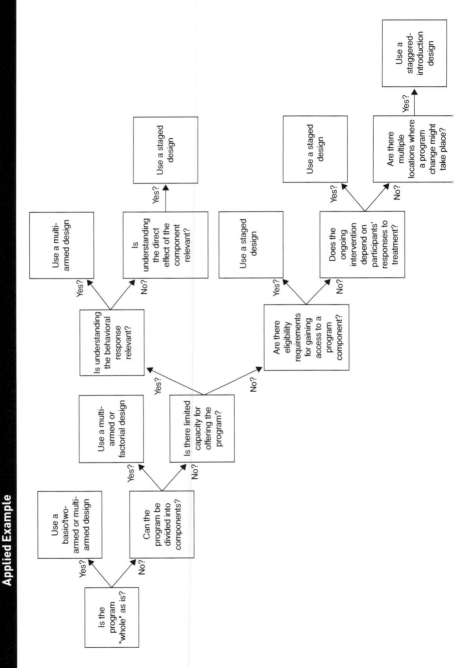

(Continued from p. 42)

of the program component to the program's overall effectiveness? To undertake the exercise of aligning certain program characteristics with evaluation design options, consider the following questions:

- Is the program "whole" as is (as opposed to itself comprising multiple components)?

- Can the program be divided into components (and does the research question imply that looking at it in its component pieces is relevant)?

- Is there limited capacity for offering the program or a given component of it?

- Are there additional eligibility requirements or triggers (including the passage of time) that screen who should gain access to a program component?

- Does the ongoing intervention design depend on participants' initial responses to treatment?

- Are there multiple locations where a program change might take place?

These questions are about the specific characteristics of a program, where the answers would provide input to managers' program improvement efforts. Figure 4.4 places these questions into a decision tree, where—depending on the answer—the particular experimental evaluation design is identified as fitting.

Applied Example. This section draws examples from career pathways initiatives to illustrate which program characteristics might be examined using which designs to help answer specific research questions. The reason that career pathways initiatives are a useful example is that they are, by definition, multifaceted interventions. Career pathways programs are customized to specific participants' needs and interests, and they include varied program components—education, training, support services, employer connections—to carry out their aim of helping people enter onto a career path and not just into a job. The impact of these collective efforts is of interest, but so too is understanding what about the multifaceted approach matters for both policy and practice. As such, career pathways programs provide an ideal environment in which to consider how to connect specific program components and questions about their relative impacts to appropriate evaluation designs.

Table 4.3 offers some examples from the career pathways program world that illustrate how to think about aligning certain program component characteristics with evaluation designs to reveal the relative contributions of those program components. The four kinds of program models or components that appear in the left-most column of the exhibit are Blended Learning Plus, Emergency Assistance, On-the-job

BOX 4.1
EXAMPLES OF COMPONENTS IN CAREER PATHWAYS TRAINING PROGRAMS

- *Blended learning plus (BLP)* offers students a combination of face-to-face instruction and online learning (the blended learning), plus access to a laptop and reliable, fast, in-home Internet access. Aspects of BLP include the instructional mode, students' hardware, and students' Internet access. These three aspects could be considered together or separately.

- *Emergency assistance* is a support where individuals who experience an emergency of sorts are given assistance so that their circumstances do not interfere with their program participation. Programs might offer this assistance for vehicle repair, public transit subsidy, eviction prevention, utilities assistance, uninsured medical emergencies, child care, and food.

- *On-the-job training (OJT)* is training that an employer (in the public, private nonprofit, or private for-profit sector) offers to a participant, as part of employment in a job for a limited time. In the context of HPOG, OJT is available through the grant or a federally-funded program that provides partial reimbursement to the employer for participant wages to offset costs of training. Generally, OJT slots are limited in number and limited to trainees who have met certain program milestones.

- *Mandatory work-readiness* as a component of a career pathways program involves required intensive work-readiness training; *peer support* is a component that brings participants together into cohort groups where they can gain work-related interpersonal, intrapersonal, and workplace navigation skills. While both work readiness and peer support exist in the career pathways programs, prior evidence—both empirical and anecdotal—suggests that combining them might confer additional benefits to trainees. So, for this example, the component examined is the pair of these individual components, combined together.

Training, and Mandatory Work-Readiness with Peer Support. These were program innovations considered as part of the development of the second round of HPOG grant funding and the evaluation of that grant program (Harvill et al., 2015; Peck, 2016a). Each of these is defined in more detail in the textbox.

Consider the BLP program component. If the program component as a whole is what is of interest to policy and practice, then the design should randomize treated individuals to gain access to BLP as a whole. But if the distinct contribution of giving students computers is of interest independent from BLP's instructional practice, then using a factorial design would be appropriate. A factorial evaluation design would involve randomizing individuals to be part of a standard career pathways program

TABLE 4.3 ■	Characteristics of Some Career Pathways Program Components and Implications for Evaluation Design	
Program Component	**Characteristics of Program Component**	**Implications for Evaluation Design**
Blended Learning Plus	• Includes three "pieces": (1) Instruction mode: part online (2) Student hardware access (3) Internet connection at home • Can be considered as a whole or in parts	Multi-faceted component could be considered: • as a whole → basic design • in pieces → factorial design
Emergency Assistance	• Provides assistance to participants to prevent drop-out associated with financial crisis (such as eviction, car breakdown) • Involves an additional eligibility requirement • May be limited in amount of funding (overall and/or per person)	• Limited availability → ration access through (1) added treatment arm, or (2) second stage randomization • Additional eligibility requirement → staged design • Response to treatment determines course of treatment → staged design
On-the-job Training	• Includes subsidy to employers to permit trainees to work while training • Involves an additional eligibility requirement • Has limited slots	• Limited availability → ration access through (1) added treatment arm or (2) second stage randomization • Additional eligibility requirement → staged design • Response to treatment determines course of treatment → staged design
Mandatory Work-Readiness With Peer Support	• Involves (1) work-related skills development and (2) cohort-based peer support • Can be considered as a whole or in parts	Two-pronged component could be considered: • as a whole → basic design • in pieces → multi-armed or factorial design

(Cell 1, from Figure 4.1 or 4.2), in which individuals gain access to the program without any of the added aspects of BLP. The design would also randomize individuals be part of the program while also having the opportunity to take their courses via the online/blended instructional mode (Cell 2), to be given a computer for use during their training and a subsidy for at-home Internet service (Cell 3), or to have both online/blended instruction and a computer/home Internet access (Cell 4). Like BLP, the Mandatory work-Readiness with Peer Support program component can allowing combined or separate evaluation of multiple subcomponents, depending on the evaluation question(s) of interest.

Next, program components with limited enrollment capacity are candidates for a multi-armed evaluation design, where the program's capacity is used to determine the number or proportion of cases that the study would randomize into that component. For example, if the program has funding of $50,000 to provide emergency assistance in response to participants' needs, then giving people access to those resources via a lottery is a fair way to distribute those funds while also providing an opportunity to learn about the effectiveness of such assistance. This can occur through a three-armed experiment where the second treatment arm offers individuals access to the program with the possibility of emergency assistance if for those individuals who need it (such as having to pay for car repairs in order to help them get to work); or it could occur through a staged evaluation design where those individuals who have an emergency need can enter a lottery to access those funds. The size of either enhanced treatment group (either access from the beginning or access once there is demonstrated need) can accommodate the amount of support budgeted.

Because OJT capacity is limited, offering OJT could be fairly distributed via a lottery that is open only to trainees who have met certain program milestones. This additional eligibility requirement implies a staged design, where eligibility triggers potential access.

To further elaborate on the emergency assistance enhancement that is part of the first grant cycle of HPOG: the HPOG Impact Study randomizes people into a second treatment arm where they are offered access to emergency assistance should they need it. Fortunately for the aggregate well-being of the treatment group, only about a third of individuals had emergencies that triggered their use of that assistance (Peck et al., 2018a). As such, the impact of the HPOG emergency assistance enhancement will be of the *offer* of assistance, where need/take-up is modest. This design option captures behavioral response: If people know they have a safety net, they may behave accordingly. In the case of a vehicle breakdown, instead of taking the bus or asking others for a ride, they could access the program's assistance in order to fix the vehicle, a cost they might not otherwise have incurred (or incurred as soon).

In comparison, if researchers or policymakers would want to know the direct effect of receiving assistance, then a staged design should be used. For example, once an individual triggers the need for assistance, she would be randomized to receive that assistance or not. Given that we cannot afford to help everyone who has an emergency, the fairest way to do so is via a lottery. Either way—through a multi-armed design or a staged design—some individuals will be randomized out of having access to emergency assistance, but the immediacy of the need is closer to randomization in the case of the staged design than it is with a multi-armed design. As such, the latter might seem more palatable to program staff who have to communicate the results of the lottery to their program participants. However, if we want to know the effect of receiving that component, then a staged design is fitting.

Another career pathways program component is labeled Mandatory Work-Readiness with Peer Support. As noted above, each of these program components currently exists in the field, and anecdotal evidence suggests their potential effectiveness independent of one another and in combination. In order to ascertain whether these two program components have independent and synergistic impacts, a factorial experimental design could be ideal. An excluded control group is not needed. Instead, from the standard program some (randomly assigned) individuals would have the work-readiness training requirement, others would have peer support, and others still would engage in both. If programs are aiming to identify how to enhance their offerings in multiple ways, then a factorial experimental design can provide the evidence needed to support decisions about whether that expansion should involve one or both of two such enhancements.

Considering these career pathways program examples, it is clear that a given "component" can itself be multifaceted. For example, the BLP example is itself multifaceted and therefore a candidate for multi-armed or factorial designs. The specific evaluation of interest will determine the specific design that is most appropriate. Similarly, OJT is an example of a program component where an earlier program milestone must be met in order for participants to access. Many other kinds of programs lack the resources to offer all components to every participant. Moreover, programs are often evolving, and as they evolve can consider how to test new components that might improve the overall program's effectiveness.

CONCLUSION

This chapter has identified several variants of experimental evaluation designs that can be used to examine specific aspects of a program's effectiveness. Randomization is a powerful tool for eliminating alternative explanations for why impacts arise. As Manzi (2012) encourages and foreshadows, an "experimental culture" has the potential to benefit government and society. The ideas I have presented aim to highlight how flexible experimental designs can be, that they need not oppose ethics or compassionate practice, and that program administrators and funders should embrace an experimental culture in order to maximize what we learn and the quality of what we learn about evolving programs to achieve better outcomes.

QUESTIONS AND EXERCISES

1. Identify a program design or administrative challenge, and use one or more of the evaluation designs described in this chapter to establish a strategy for learning how program changes might improve outcomes.

2. Consider a program that you are familiar with and propose a blended design that combines an enhanced treatment and a staged design to evaluate certain aspects of that program.

3. Discuss criteria for choosing each design. Consider a program that you are familiar with and discuss the tradeoffs of using one design over another, given possible questions of relevance to possible program changes.

4. Identify the program components of a program you are familiar with. Given the characteristics of each component, identify which experimental evaluation design lends itself to assessing the effectiveness of each.

5. Think of a program you know that lacks resources to deliver all services or all components to everyone. How would you design an evaluation that might allocate services in a way to learn about their relative effectiveness?

5 PRACTICAL CONSIDERATIONS AND CONCLUSION

Too often, researchers and practitioners believe that an experimental control group is denied services. This is not the case. Experimental evaluation designs need not involve an excluded control group. They can be a powerful tool for learning about the impacts both of programs overall as well as the component pieces of programs. Policy experiments can follow the lead of the private sector and increasingly use experiments for assessing which program models or various program designs are most effective. Using experiments for program learning and program improvement efforts is both feasible and practical.

This concluding chapter first raises some issues that researchers and practitioners will encounter as they implement these experimental designs in practice. Then it discusses the concept of road testing, a practitioner's approach to establishing an organization that uses rigorous evaluation input on a regular basis. Finally, it suggests a set of principles that rigorous evaluations should follow.

SOME PRACTICAL CONSIDERATIONS

This book provides an introduction to experiments, both as commonly understood and in flexible variants that can support program improvement efforts. A rule of thumb for evaluation is that it should be funded at the level of about 10% of a program's overall cost. This can occur by contracting with an outside, third-party evaluator or by having a designated staff member or team internal to the organization facilitate and help carry out the work.

Regardless, implementing any of these designs in practice will require careful planning and may require input from an expert evaluator. That said, *do* try this at home! I encourage practitioners interested in learning more about what drive their programs' impacts to imagine how to embed randomization into standard program processes in order to permit thoughtful evaluation.

Sample Size Considerations

Commonly, when an impact evaluation is in its design or planning phase, the evaluator will undertake what is called a "power analysis" to determine the ideal number of study units needed. The size of the impact that has relevance for policy or practice determines the number of observations that should be part of an evaluation.

Box 5.1 discusses the factors that determine a detectible effect sizes. Generally, the larger the sample, the smaller the effect size that a study can detect. In practice, however, achieving small effects is rarely either a program goal or of practical interest. Instead, program innovations that make a big difference to outcomes are the ones that warrant being pursued. To answer questions about whether some program change will result in large changes to program outcomes may require only a modest sample size. Regardless, advance planning will help identify what an ideal sample size would be for a given impact size.

Random Assignment in Practice

The point at which research units are randomized has implications for subsequent interpretation of evaluation results. A general rule of thumb is that the point of random assignment is the place within a program where impact results can be generalized to. For example, if individuals eligible for the program are randomly assigned before they experience any of the program, then the impact is attributed to the whole program. In comparison, if using a staged design, when random assignment takes place after a program participant achieves some milestone that earns him or her an additional program opportunity, then the comparison of the second-stage outcomes offer evidence on the effectiveness of that second-stage opportunity.

Once the right point within a program flow is identified, then the practical mechanism for implementing random assignment should be chosen. Random assignment can be embedded into existing program management information systems or can be stand-alone software interfaces. Again, these are commonly put into place via a partnership with a professional evaluator. As Gifford-Hawkins explains (see Box 5.1), once a program becomes familiar with evaluation procedures, its own organizational capacity may be sufficient to implement them independently.

What Local Programs Should Know About Generalizability

The experimental designs detailed in this book emphasize **internal validity**, or an evaluation's ability to establish a causal connection between a program and impacts. This is appropriate because any given program will be most interested in what the impacts of its own efforts are on its own program participants. That information, however, may not be generalizable to other programs, in other settings, serving different target populations, or at different times. The generalizability of an evaluation's results is also referred to as the evaluation's **external validity**.

In recent years, scholars and practitioners alike have increased their attention on how experiments can provide results that have greater generalizability. This is of value because it has the potential to increase what program administrators' can learn, if not from their programs then from other, similar programs. As consumers of evaluation results, local program administrators should pay attention to program designs, settings, targets, and context in order to judge how generalizable results are to them.

BOX 5.1

HOW LARIMER COUNTY EMBRACES EXPERIMENTATION

By Ella Gifford-Hawkins

The Larimer County (Colorado) Economic and Workforce Development Department Works Team is committed to fostering a work culture that embraces continuous learning. When we undertake any change to improve our program, we deliberately "road test" it before rolling out the change. A road test is an analytic approach to piloting changes on a small scale to ensure that the design and implementation are right before scaling up. Road tests provide us with timely evidence about how the change is playing out for staff members and program participants and allow us to strengthen our implementation approach. If the road test does not show early signs of promise, then we correct course early on, saving time, money, and effort. Two of the road tests we have undertaken to date include the following:

- **Goal Achievement Coaching.** This particular road test took place over a 6-week period with weekly surveys sent to each coach to gather feedback on the approach and tools. This road test helped us identify gaps in staff training and support necessary to sustain the coaching model; as a result of the road test, we instituted practice sessions and other reinforcement opportunities that increased coaches' comfort with the new model. Goal achievement coaching is now integrated within our service delivery process.

- **Monthly Time Sheet (MTS) Road Test.** Obtaining the required participant monthly time sheets, which record participants' participation in required program activities, had long been a challenge for our team. We thought that changing the time sheet into an electronic form accessible through our online platform, and providing it in a mobile-friendly format, might be a good solution. We wanted to learn how the MTS functionality in the online platform, and a mobile-friendly format, could make it easier for participants to submit their required monthly participation information and how it could also reduce the administrative burden on staff associated with MTS compliance and verification. Road test feedback from coaches and participants indicated that most, but not all participants, were able to submit the time sheet easily and more timely. We continue to refine and improve this functionality through road tests.

Our team's use of road tests, and the resulting change in our organizational culture, is a product of our ongoing partnership with Mathematica Policy Research, a research and technical assistance firm. The Mathematica team helped us adopt and employ the *Learn, Innovate, Improve* (LI^2) framework to identify program inefficiencies, generate innovative approaches to improve outcomes, and deploy research methods for testing new approaches. Now that we have learned and embedded these capabilities in-house, road testing is a standard part of our organizational life. Before making a change, we start with a road test.

(Continued)

(Continued)

Road tests provide structure, timelines, and a process to learn and make improvements and scale up change. In addition, staff not only see the agency making decisions and taking action, they play an active role in the change process and provide feedback, insight, and information based on their experience. Customers are also actively involved in many road tests and we have found their feedback to be invaluable in our ability to enhance customer service and program improvement efforts.

In the past, we have made purchases based on instinct and what we thought might be needed. Now, we use road tests to systematically gather information on a proposed change to inform our decision making.

ROAD TESTING

Recent increased emphasis on becoming a "learning" organization demands that program managers embrace evidence to inform their decisions about undertaking program change efforts. In practice, this means that organizations are becoming increasingly open to "road test" experiences, where they implement change deliberately and assess the effects of those efforts to decide whether to continue or scale up the changes. That field-based terminology is in sync with the experimental evaluation design options detailed in this book. The practice of road testing via deliberate experimentation can be more widely understood and implemented across the public and nonprofit sectors as a means to inform program improvement efforts. With increasing media attention on experiments and their benefits in the business sector, it is my hope that a culture of experimentation can evolve and improve the quality of evidence we use to establish and improve public and nonprofit programs in practice.

PRINCIPLES FOR CONDUCTING HIGH-QUALITY EVALUATION

This section proposes a set of 20 principles that should be followed for conducting high-quality experiments in practice (see Box 5.2). These were developed by my Abt Associates colleagues Randall Juras and Beth Boulay based on their and the company's collective experience operating experimental evaluations generally and providing technical assistance to more than 200 rigorous evaluations, mostly funded by the U.S. Departments of Education and Labor. Although some of the principles are specific to experiments (e.g., those items that pertain to the process and implications of randomization), other principles are relevant to good practice in impact evaluations more broadly. Regardless, the principles are practical considerations for carrying out a high-quality randomized experiment that I believe all impact evaluators should follow.

BOX 5.2
PRINCIPLES FOR CONDUCTING EXPERIMENTS IN PRACTICE

1. Randomized experiments should ensure the protection of human subjects and the confidentiality of their personal information.

2. Randomized experiments should be guided by scientific principles and insulated from political influences.

3. Randomized experiments should provide a description of the intervention under investigation and measure and report the fidelity with which the intervention was implemented.

4. Randomized experiments should clearly describe the control or counterfactual condition(s) to which the intervention is being compared.

5. Randomized experiments should specify clear research questions that state the intervention, the counterfactual, the outcome domain, and the research participant sample.

6. Randomized experiments should use an appropriate power analysis to ensure that the planned design includes an adequately sized sample and can detect meaningful impacts (as defined by policy or practice) of the intervention with high probability.

7. Randomized experiments should have a plan to recruit participants—including securing their informed consent to participate—that is well integrated into the ongoing activities of the host organization and the context in which the intervention is being implemented. The recruitment plan should include materials that make the evaluation goals and requirements clear to a wide range of stakeholders.

8. Randomized experiments should articulate how research participants will be sampled and randomly assigned to study conditions and describe the randomization mechanism and procedures, and attest to their reliability.

9. Randomized experiments should have a clearly articulated and feasible timeline, including plans for how long research participants will be followed and whether multiple cohorts of research participants will be included. A discussion of any possible spillovers, contamination, and general equilibrium effects should be addressed in the design.

10. Randomized experiments should track research participants over time from random assignment to outcome data collection and analysis. Evaluators should document and report the extent of and reasons for follow-up data collection sample loss (attrition) by study conditions, including examining (and attempting to mitigate) its consequences in creating potential impact estimate bias.

(Continued)

(Continued)

11. Randomized experiments should track the number of control group members who receive the intervention (crossovers) and the number of treatment group members who do not receive the intervention (no-shows).

12. Randomized experiments should have a well-articulated data collection plan and process for tracking and documenting data collection as the evaluation is conducted.

13. Randomized experiments should include a detailed plan for assessing data quality, including treatment of outliers or implausible values, and use of any validation data.

14. Randomized experiments should have a well-articulated plan for storing and managing raw data and analytic data files and for documenting what analyses are conducted and where analytic programs and results are stored.

15. Randomized experiments should include a detailed plan for analysis of impacts, including detailed model specifications; a plan for addressing multiple comparisons to avoid an unduly high rate of false positive findings; identification of **subgroups** for separate analysis; inclusion/exclusion rules for covariates; and treatment of missing data, including possible imputation methods and/or nonresponse weighting.

16. Randomized experiments should include plans to obtain valid estimates of the variability of impact estimates, with special attention to reliable standard errors.

17. Randomized experiments should attend to the study's external validity—that is, the ability to generalize results to a broader population of interest—and address policy implications where feasible. Caveats to interpretation should be stated in clear, accessible language.

18. Randomized experiments should provide clear and transparent reports of estimated impacts, with all relevant associated statistics (e.g., tests of statistical significance), and use clear means of communicating those results (e.g., tables or graphics).

19. Impact evaluation reports and other dissemination products should be written in clear language that is accessible to a nontechnical audience (excepting methodological appendices).

20. Evaluation documents should undergo thorough quality review by an expert or experts (outside the core evaluation team) prior to submission to the client and/or publication.

These principles provide a framework for evaluators, administrators interested in evaluating their programs or aspects of their programs, and funders of evaluation to employ to ensure the high quality of their work. These practical principles align with the American Evaluation Association's "Guiding Principles" (AEA, n.d.) that evaluators undertake their work with systematic inquiry (evaluators conduct systematic, data-based inquiries), competence (evaluators provide competent performance to stakeholders), honesty and integrity (evaluators display honesty and integrity in their own behavior, and attempt to ensure the honesty and integrity of the entire evaluation process), respect for people (evaluators respect the security, dignity, and self-worth of respondents, program participants, clients, and other evaluation stakeholders), and responsibilities for general and public welfare (evaluators articulate and take into account the diversity of general and public interests and values that may be related to the evaluation).

In summary, this book has presented a variety of experimental evaluation designs that highlight how experiments can offer insights into questions about *what about* a program drives its impacts. To date, these questions have been thought to fall under the auspices of nonexperimental evaluation designs. Instead, if we consider randomization across multiple treatments, across multiple treatment components, and across stages of a program process, we can see opportunities for embedding experimentation into program operations in order to inform how best to develop and improve programs in the field.

QUESTIONS AND EXERCISES

1. Consider an evaluation you know and love. Review that evaluation's performance along the 20 principles suggested in this chapter. Do you still love it? If so, why? If not, what are its noticeable shortcomings?

RESOURCES FOR ADDITIONAL LEARNING

*   American Evaluation Association's "Find An Evaluator" service: https://www .eval.org/p/cm/ld/fid=108

*   Research Randomizer: https://www.randomizer.org/

APPENDIX

Doing the Math and Other Technical Considerations

As noted, this book is about evaluation design variants that can help answer questions about the impacts of program improvement efforts. That said, it is still essential to understand the basic math associated with carrying out an impact analysis. Fortunately, only simple subtraction is strictly necessary in order to estimate program impacts: the treatment group's mean outcome minus the control group's mean outcome equals the impact of the program. In addition to providing an introduction to the analysis of experimental data, this appendix will cover some other technical considerations that will be helpful to analysts, such as the following:

- interpreting results, including **statistical significance** and **null results**;

- handling treatment group "no-shows" and control group "crossovers";

- sample size considerations; and

- basic **subgroup analyses**.

ESTIMATING TREATMENT IMPACTS

This section provides basic estimation procedures for the evaluation designs discussed in Chapters 3 and 4. To begin, an experimental evaluation's estimate of the impact is computed as the difference between the average outcome of the treatment group and the average outcome of the control group. This can be written as follows:

$$\Delta = \bar{Y}_T - \bar{Y}_C$$

Delta, Δ, refers to the change in mean outcomes, or the "impact." Y-bar is the mean outcome for each of the groups of interest, subscripted here as T for treatment and C for control. This math is simple enough that a calculator is barely needed to carry it out. For any evaluation contrast, the mean outcome of the control group (or status quo treatment arm) is netted out of the mean outcome for the treatment (or alternative treatment) group. This basic concept can be extended across most designs: A comparison of the two values of interest produces an estimate of the impact, or difference, between the two.

More conventionally in practice, impacts are analyzed in a multivariate regression framework. The reason for this is that data collected at baseline can be used to control for differences between the treatment and control groups' characteristics. Although random assignment assures that there are no systematic differences between the groups (and this is increasingly the case as samples get larger), inevitably there will be "noise." That noise reduces the analyst's ability to detect effects; and so, if it can be controlled for, then the analyst can be more precise in estimating program impacts. To carry this out in practice, the standard regression equation for a basic (two-armed) experimental evaluation is specified as follows:

$$Y = \alpha + \delta T + \beta X + e$$

In this equation, Y is the outcome of interest. T is a dummy variable that indicates assignment to the treatment group (those in the control group assume a value of 0). X is a set of individual background characteristics, and e is a random error term. The intercept, α, represents the regression-adjusted mean outcome for the status quo (control) group. The coefficient, δ (or delta), is interpreted as the impact of being in the treatment group, which is known as the ITT ("intent to treat") estimator. The β coefficients noted above are not of substantive interest.

To analyze data from a three-armed randomized experimental design, the basic equation adds an indicator for the second treatment arm, as follows:

$$Y = \alpha + \delta_1 T_1 + \delta_2 T_2 + \beta X + e$$

As above, Y remains the outcome of interest. Here, T_1 is a dummy variable that indicates assignment to one of the treatment groups, and T_2 represents assignment to a second treatment arm. Those in the control group carry a value of 0 for both Ts. Again, the intercept, α, represents the regression-adjusted mean outcome for the control group. The coefficient δ_1 is interpreted as the impact of being in treatment arm 1, and δ_2 is interpreted as the impact of being in treatment arm 2. The β coefficients noted above are not of substantive interest. In the three-armed analysis, one can test the hypothesis that $\delta_1 = \delta_2$.

Next, a 2x2 randomized factorial design permits analysis of eight questions. As noted, the power of the factorial design comes from its use of those randomized not to receive Factor A as the control group for comparing the outcomes of those assigned to Factor A, for example. This is what is represented in the "overall" impact estimate. This is in contrast to the impact of Factor A "alone," which uses only the control group's mean outcomes as the counterfactual. The general regression model associated with this 2x2 factorial design is as follows:

$$y = \alpha + \delta_A F_A + \delta_B F_B + \delta_{AB} F_A F_B + \beta X + e$$

The outcome of interest is y. F_A is a dummy variable for the presence of Factor A (or of high-intensity Factor A) in the assigned intervention (= 1 when Factor A is present, = 0 otherwise), and F_B is a dummy variable for the presence of Factor B (or of high-intensity Factor B) in the assigned intervention (= 1 when Factor B is present, = 0 otherwise). As with the previously specified models, α, the intercept, remains interpreted as the mean control group outcome. One can add baseline covariates (the Xs), which serves to increase the precision of the impact estimates, as in a conventional impact estimating model; and, again, interpreting their coefficients is generally not of interest. Most centrally, the δs represent the impact of their corresponding factors: δ_A is the mean impact of factor A; δ_B is the mean impact of factor B; and δ_{AB} is the added synergistic effect of combining Factors A and B in addition to the simple additive contribution.

HOW TO INTERPRET RESULTS

As noted in the discussion above, the coefficient on the treatment term of interest (where that term is coded as 1 for the treatment group and 0 for the control group) is interpreted as the impact of being in the treatment group. So if the outcome examined is dollars and the coefficient is 137, then the interpretation would be as follows: The treatment group earned $137 more than the control group.

An aspect of this interpretation also is whether that value is statistically distinguishable from zero. In the case that the analyst is comparing two mean values, a t-test on the difference in means will ascertain whether the difference is statistically distinguishable from zero. In the context of multivariate regression, the analysis also uses a t-test to assess whether the coefficient differs from zero. The probability (p value) associated with whether the value of the t-statistic places it outside the critical range reflects the analyst's threshold for establishing statistical significance. Commonly, evaluation reports use asterisks to identify whether a result is statistically significant at the $p < 0.01$, $p < 0.05$, and $p < 0.10$ levels, as shown in the sample results table in Table A.1. The threshold for statistical significance guards against false positive results.

The elements that appear in Table A.1 either come from or are computed from the regression output. As noted above, the value for the control group mean ($3,345) is the estimating model's intercept (α), and the impact value ($137) is the coefficient

TABLE A.1 ■ Sample Table of Results

Outcome	Treatment Group Mean	Control Group Mean	Difference (Impact)	Relative Impact
Earnings in Quarter 5	$3,482	$3,345	$137**	4.1%

Notes: Due to rounding, reported impacts (T-C differences) may differ from differences between reported regression-adjusted means for the treatment and control groups. Sample includes study participants with non-missing outcome data.

*** Difference is statistically significant (two-sided t-test) at the p < 0.01 level.

** Difference is statistically significant (two-sided t-test) at the p < 0.05 level.

* Difference is statistically significant (two-sided t-test) at the p < 0.10 level.

on the treatment indicator, δ. The treatment group mean (\$3,482) is computed by adding together the values of α and δ. And the relative impact is computed by dividing the impact value by the control group mean. Although this latter value is not always included in evaluation reports, I find it to be important as part of interpreting the magnitude of the impact. Is \$137 a big or a small number? Relative to the control group mean, that absolute value has a relative size of 4.1%. The other part of interpreting the size of the impact magnitude involves relating it to other research findings: Is \$137 smaller, roughly the same, or larger than other evaluations have observed as a quarterly earnings impact for this kind of program?

Interpreting Null Results and the Role of Sample Size

It is common for evaluations to fail to detect statistically significant results. Indeed, years of experimentation have found that—at least in the educational and social policy arenas—it is difficult to identify interventions that are strong enough to produce large effects. Peter Rossi's "iron law" states that "the expected value of any net impact assessment of any large scale program is zero" (1987, p. 4), which he follows by noting (in the "stainless steel law") that better designed impact assessments are associated with the even greater likelihood that the detected impact will be zero, and (in the "zinc law") "only those programs that are likely to fail are evaluated" (Rossi, 1987, p. 5). Although those laws were not based on any empirical evidence (Rossi, 2003), they are often compared to observations from practice. Indeed, what might characterize the past three decades of research on job training programs, for example, is that they are marked by null or small impacts (e.g., Haskins & Margolis, 2014). While policy innovators are working on what kinds of program design options might generate larger impacts—perhaps in the face of Rossi's (1987) "brass law" that "the more social programs are designed to change individuals, the more likely the net impact of the program will be zero" (p. 5)—we must still consider the likelihood that evaluations will not reveal large and statistically significant program impacts. It is this point that I would like to elaborate in providing some guidance for how to interpret what might seem like the "null" effect of a program.

To do so, the concept of the **minimum detectable effect (MDE)** is helpful (see Bloom 1995a, 1995b). The MDE is the smallest true intervention impact that can be detected, under the given empirical circumstances (including the sample size, data and measures, and statistical assumptions). Often projects conduct a "power analysis" and use MDEs in planning to ensure that the target sample size will be sufficient to detect an effect of a magnitude that is relevant to policy or administration. I have also found that computing MDEs after the fact can be very helpful because they can identify how large the impact *would have needed* to be to be detected under the project's circumstances. MDEs are a function of a variety of factors, which are summarized in Box A.1.

BOX A.1
FACTORS THAT DETERMINE MINIMUM DETECTIBLE EFFECT SIZES

- *Statistical significance threshold.* The statistical significance level is the probability of identifying a **false positive** result (also referred to as **Type I error**). The MDE becomes larger as the statistical significance level decreases. All else equal, an impact must be larger to be detected with a statistical significance threshold of 1% than with a statistical significance threshold of 10%. The decision about choosing a statistical significance threshold hinges on the analyst's or administrator's tolerance or funder's cost tolerance for having to live with a potentially false positive result.

- *Statistical power.* The statistical power is equal to the probability of correctly rejecting the null hypothesis (or, one minus the probability of a **false negative** result, or **Type II error.**) In other words, power relates to the analyst's ability to detect an impact, should it exist. Statistical power is typically set to 80%, but there is no reason that one could not increase (say, to 90%) or decrease (say, to 70%) that value, again depending on the confidence with which one wants to place on a given result. Missing the detection of a favorable impact (Type II error) has lower cost implications, relative to falsely claiming that a favorable impact exists (Type I error).

- *Variance of the impact estimate.* Variance is essentially a measure of the "noisiness" of the impact estimate (the precision/imprecision with which the impact is measured). Some kinds of measures, such as incomes within a diverse population, are inherently noisier and therefore more difficult to detect systematic and small impacts. The magnitude of a detectable effect increases as the variance of the impact estimate increases. The variance of the impact estimate is a function of sample size, with larger samples producing lower variance. In turn, the MDE will also be influenced by sample size, such that MDEs will increase as sample size decreases.

- *Magnitude of the control mean.* For outcomes that are very infrequently observed or small in magnitude, achieving a *change* in the treatment group mean will be detectable only if that change is very small in absolute magnitude, which may still imply a large relative impact. For example, if just 1% of the control group in a homebuyer financial management intervention defaults on their mortgages, then—in order for a detectable effect to arise, given statistical significance threshold, power, variance, and sample size—the magnitude of the effect will have to be large relatively. If the MDE is 1 percentage point, that implies a 100% relative effect, or a reduction to zero for the treatment group relative to the control group mean (of one percent). That seems unlikely to arise. In comparison, if the control group mean were 8% and the MDE were still 1 percentage point, then that magnitude change would reflect a 12.5% relative impact, which seems more likely to arise. In brief, "small" effects are simply harder to detect on a small base.

I have discussed these issue of the MDE as a way to help interpret results that are not identified as statistically significant. In situations where impacts are not statistically significantly different from zero, it does not necessarily mean that there are "no impacts." Rather, it means that there are "no *detectable* impacts." This semantic distinction is important to make. There may in fact be impacts that are smaller than the circumstances of a given study can detect, and it would be irresponsible to label a program "ineffective" (which implies a precisely measured zero impact) when the program does not have a large impact (at least of the magnitude detectable under the study's circumstances). Computing a post hoc MDE is a way to place a value on the magnitude of the impact that would have needed to arise to be detectable under the study's circumstances. In recent years, the practice of computing post hoc MDE is becoming more commonplace (e.g., Harvill et al., 2018; Peck et al., 2019).

HANDLING TREATMENT GROUP NO-SHOWS AND CONTROL GROUP CROSSOVERS

The main discussion of impact estimation in this chapter involves what is called the **intent-to-treat** (or ITT) impact. This impact estimate involves comparing the whole treatment group to the whole control group, regardless whether any of the treatment group members were **no-shows**, and regardless of whether any of the control group members finagled their way into treatment and were **crossovers**. An alternative impact estimate, called the **treatment on the treated** (or TOT), involves making some assumptions in order to compute an impact for those who were actually treated. Sometimes evaluations where no-shows are expected are called "encouragement" designs, in that they encourage, but cannot force, treatment group members to take up the offer. The resulting ITT impact estimate includes the whole treatment group, including those who took up the offer and those who did not. As such, it captures the impact of the behavioral response to the offer as well as the impact of engaging in the treatment offered. Commonly, the ITT is considered the more policy relevant impact because it assesses the effect of making some policy or program available but not forcing people to participate. From a program administrator's perspective, however, the TOT is probably the more relevant because it assesses the program's effect for those who show up to participate.

The *pattern of results* is the same whether one considers the ITT or the TOT impacts, and only the specific point estimates differ in magnitude. As a result, consumers of evaluation research that reports both ITT and TOT results can choose which type of impact is more relevant to their interests.

Although adjusting for no-shows and crossovers can be executed analytically using a two-stage least squares multiple regression procedure, the underlying math—just like that of the basic impact analysis—is intuitively simple. If we can assume that there is no impact on those who are no-shows, then the ITT impact can be spread across those who are participants. Mathematically, the expression for this is as follows:

$$ITT = \frac{TOT}{(treatment\ receipt\ rate)}$$

Very often control group members are strictly excluded from accessing program services, but there are some circumstances under which they might be able to gain access. In this situation, if we can assume that the impact on control group crossovers is the same as the impact on treatment group members who took up, then the ITT impact can be recalibrated to represent the TOT impact, as follows (e.g., Bloom, 2006):

$$ITT = \frac{TOT}{(treatment\ receipt\ rate - control\ crossover\ rate)}$$

In both instances, the "rescaling" of the ITT estimate to represent a TOT estimate results in both estimates sharing the same level of statistical significance (Bloom, 2006). That is when an ITT estimate is statistically significant, then its corresponding TOT estimate is statistically significant as well. When the ITT estimate is not statistically significant, its corresponding TOT estimate also is not statistically significant. In other words, the pattern of results is the same for both ITT and TOT impacts, but the magnitude differs based on the underlying assumptions. Litwok and Peck (2018) demonstrate that this is likely the case in theory, although it might not be in practice, due to some technical issues and assumptions, under some circumstances.

Of course, in situations where the basic assumptions are not viable, converting an ITT estimate to a TOT estimate is not warranted. For example, in the welfare reform experiments of the 1980s and 1990s, a new policy regime imposed sanctions on people for not complying with program rules. Those sanctions took effect specifically for people who did not show up. Therefore, there could be no "no-shows" in terms of their subjection to the sanction policy.

Those who take up treatment when offered are in reality a subgroup. More often, researchers are interested in different subgroups, namely those defined by population characteristics.

SUBGROUP ANALYSES

Any subgroup that is defined at baseline is a candidate for separate analysis. A baseline characteristic is one that is exogenous to the treatment and otherwise cannot be manipulated by treatment assignment. Often researchers, program administrators and policymakers are interested to know whether an intervention has differential effects for various groups of individuals. For example, is the program more effective for women than for men? Is the program more effective for those who come into it with little (or relatively greater) work experience? These kinds of subgroup characteristics are referred to as **moderators**—that is, these characteristics moderate program impacts by influencing them through interaction effects.

This kind of analysis can take place in one of two straightforward ways: First, a **subgroup analysis, split sample** involves dividing the sample (both treatment and control groups) into distinct groups as defined by their baseline traits (e.g., men in one group, and women in another); and then executing the impact analysis appropriate to that design, as described earlier in this appendix. Alternatively, the sample can be kept intact, and an "interaction" term can be added to the estimating equation. This **subgroup analysis, interacted model** is specified as follows:

$$Y = \alpha + \delta_1 T + \delta_2 TS + \beta X + e$$

In this equation, Y is the outcome of interest. T is a dummy variable that indicates assignment to the treatment group (those in the control group assume a value of 0). S is a binary indicator for the subgroup of interest, and it appears interacted with the treatment indicator. As before, X is a set of individual background characteristics, and e is a random error term. It is important to note that the set of Xs included *must* include an indicator for the subgroup being analyzed. As a result, the sum of the intercept, α, and the coefficient on the subgroup indicator represents the regression-adjusted mean outcome for the status quo (control) group. The coefficient, δ_1 (or "delta"), is interpreted as the impact of being in the treatment group; and the coefficient δ_2 represents the marginal impact of being in the subgroup.

CONCLUSION

This appendix has provided a foundation for undertaking the basic analysis of data from an experimental evaluation design. Conceptually, the impact is the difference between treatment and control outcomes; and, mathematically, it can be computed using simple subtraction or placed into a multiple regression framework. I also summarize key concepts such as minimum detectable effects (MDEs), the treatment on the treated (TOT) impact, and subgroup analysis, which are common parts of evaluation research in practice.

QUESTIONS AND EXERCISES

1. Compute the impact for the following scenario: The treatment group's mean earnings outcome is $3,250 in a given quarter, and the control group's outcome is $3,000. In addition to the absolute impact, compute the relative impact.

2. Rewrite the equation for a basic experimental impact analysis to analyze sex and age as moderators.

RESOURCES FOR ADDITIONAL LEARNING

Bloom, H. S. (1984). Accounting for no-shows in experimental evaluation designs. *Evaluation Review, 8*(2), 225–246. doi:10.1177/0193841X8400800205

Bloom, H. S. (1995a). Erratum. *Evaluation Review, 19*(5), 707. doi:10.1177/0193841X9501900607

Bloom, H. S. (1995b). Minimum detectable effects: A simple way to report the statistical power of experimental designs. *Evaluation Review, 19*(5), 547–556. doi:10.1177/0193841X9501900504

Bloom, H. S. (2006). *The core analytics of randomized experiments for social research* (MDRC Working Paper). New York, NY: MDRC. Retrieved from https://www.mdrc.org/publication/core-analytics-randomized-experiments-social-research

RCT-YES is a free software tool that permits analysis of data from an experimentally designed evaluation: https://www.rct-yes.com

REFERENCES

Aizenman, N. (2017a, August 9). Cash aid could solve poverty—But there's a catch. *Morning Edition*, National Public Radio. Retrieved from http://www.npr.org/sections/goatsandsoda/2017/08/09/542357298/cash-handouts-could-solve-poverty-but-theres-a-catch

Aizenman, N. (2017b, August 7). How to fix poverty: Why not just give people money? *All Things Considered*, National Public Radio. Retrieved from http://www.npr.org/sections/goatsandsoda/2017/08/07/541609649/how-to-fix-poverty-why-not-just-give-people-money

American Evaluation Association. (n.d.). *American Evaluation Association guiding principles for evaluators*. Retrieved from http://www.eval.org/p/cm/ld/fid=51

Babbie, E. (2016). *The basics of social research* (7th ed.). Boston, MA: Cengage Learning.

Bell, S. H., & Peck. L. R. (2016). On the feasibility of extending social experiments to wider applications. *MultiDisciplinary Journal of Evaluation, 12*(27), 93–112.

Bloom, H. S. (1984). Accounting for no-shows in experimental evaluation designs. *Evaluation Review, 8*(2), 225–246.

Bloom, H. S. (1995a). Erratum. *Evaluation Review, 19*(6), 707. doi:10.1177/0193841X9501900607

Bloom, H. S. (1995b). Minimum detectable effects: A simple way to report the statistical power of experimental designs. *Evaluation Review, 19*(5), 547–556. doi:10.1177/0193841X9501900504

Bloom, H. S. (2006). *The core analytics of randomized experiments for social research* (MDRC Working Paper). New York, NY: MDRC. Retrieved from https://www.mdrc.org/publication/core-analytics-randomized-experiments-social-research

Brooks, D. (2012, April 26). Is our adults learning? *New York Times*. Retrieved from https://www.nytimes.com/2012/04/27/opinion/brooks-is-our-adults-learning.html

Blustein, J. (2005). Toward more public discussion of the ethics of federal social program evaluation. *Journal of Policy Analysis and Management, 24*(4), 824–846. doi:10.1002/pam.20141

Collins, A., M., Briefel, R., Klerman, J. A., Rowe, G., Wolf, A., Gordon, A., . . . & Fatima, S. (2016). *Summer Electronic Benefit Transfer for Children (SEBTC) demonstration: Summary report*. Alexandria, VA: U.S. Department of Agriculture, Food and Nutrition Service, Office of Research and Analysis. Retrieved from https://www.fns.usda.gov/sites/default/files/ops/sebtcfinalreport.pdf

Collins, A., Briefel, R., Klerman, J. A., Wolf, A., Rowe, G., Logan, C., . . . & Lyskawa, J. (2015). *Summer Electronic Benefit Transfer for Children (SEBTC) demonstration: Summary report 2011–2014*. Alexandria, VA: USDA, Food and Nutrition Service, Office of Research and Analysis. Retrieved from http://www.fns.usda.gov/ops/research-and-analysis

Collins, L. M., Murphy, S., Nair, V. N., & Strecher, V. J. (2005). A strategy for optimizing and evaluating behavioral interventions. *Annals of Behavioral Medicine, 30*(1), 65–73. doi:10.1207/s15324796abm3001_8

Collins, L. M., Dziak, J. J., Kugler, K., & Trail, J. B. (2014). Factorial experiments: Efficient tools for evaluation of intervention components. *American Journal of Preventive Medicine*, *47*, 498–504. doi 10.1016/j.amepre.2014.06.021

Connell, J. P., & Kubisch, A. C. (1998). Applying a theory of change approach to the evaluation of comprehensive community initiatives: Progress, prospects, and problems. In K. Fulbright-Anderson, A. C. Kubisch, & J. P. Connell (Eds.), *New approaches to evaluating community initiatives, Vol. 2: Theory, measurement, and analysis* (pp. 15–44). Washington, DC: Aspen Institute.

Connell, J. P., Kubisch, A. C., Schorr, L. B., & Weiss, C. H. (1995). *New approaches to evaluating community initiatives: Concepts, methods, and contexts*. Washington, DC: Aspen Institute.

Greenberg, D., Meyer, R., Michalopoulos, C., & Wiseman, M. (2003). Explaining variation in the effects of welfare-to-work programs. *Evaluation Review*, *27*(4), 359–394. doi:10.1177/0193841X03254347

Greenberg, D., Meyer, R. H., & Wiseman, M. (1993). *Prying the lid from the black box: Plotting evaluation strategy for welfare employment and training programs* (Discussion Paper 999-93). Madison, WI: Institute for Research on Poverty.

Greenberg, D., Meyer, R. H., & Wiseman, M. (1994). Multisite employment and training program evaluation: A tale of three studies. *Industrial and Labor Relations Review*, *47*(4), 679–691.

Greenberg, D., & Shroder, M. (2004). *Digest of social experiments* (3rd ed.). Washington, DC: Urban Institute Press.

Gubits, D., Lin, W., Bell, S. H., & Judkins, D. (2014). *BOND implementation and evaluation: First- and second-year snapshot of earnings and benefit impacts for stage 2*. Cambridge, MA: Abt Associates. Retrieved from https://www.ssa.gov/disabilityresearch/documents/BOND_Deliverable%2024c5_8-14-14.pdf

Gubits, D., Shinn, M., Wood, M., Brown, S. R., Dastrup, S. R., & Bell, S. H. (2018). What interventions work best for families who experience homelessness? Impact estimates from the family options study. *Journal of Policy Analysis and Management*, *37*(4), 835–866. doi:10.1002/pam.22071

Gueron, J. M., & Rolston, H. (2013). *Fighting for reliable evidence*. New York, NY: Russell Sage Foundation.

Hamilton, G., Freedman, S., Gennetian, L., Michalopoulos, C., Walter, J., Adams-Ciardullo, D., & Gassman-Pines, A. (2001). *National evaluation of welfare-to-work strategies: How effective are different welfare-to-work approaches? Five-year adult and child impacts for eleven programs*. New York, NY: Manpower Demonstration Research Corporation.

Handa, S., Natali, L., Seidenfeld, D., Tembo, G., & Benjamin Davis, B. (on behalf of the Zambia Cash Transfer Evaluation Study Team). (2018). Can unconditional cash transfers raise long-term living standards? Evidence from Zambia. *Journal of Development Economics*, *133*, 42–65. doi:10.1016/j.jdeveco.2018.01.008

Harvill, E., Litwok, D., Moulton, S., Rulf Fountain A., & Peck, L. R. (2018). *Technical supplement to the health profession opportunity grants (HPOG) impact study interim report: Report appendices* (OPRE Report 2018-16b). Washington, DC: Office of Planning, Research, and Evaluation, Administration for Children and Families, U.S. Department of Health and Human Services. Retrieved from https://www.acf.hhs.gov/sites/default/files/opre/hpog_interim_appendices_final_5_15_18_508.pdf

Harvill, E. L., Sahni, S. D., Peck, L. R., & Strawn, J. (2015). *Evaluation and system design for career pathways programs: 2nd generation of HPOG (HPOG next gen design): Evaluation design recommendation report.* Washington, DC: Office of Planning, Research and Evaluation, Administration for Children and Families, U.S. Department of Health and Human Services.

Haskins, R., & Margolis, G. (2014). *Show me the evidence: Obama's fight for rigor and results in social policy.* Washington, DC: Brookings Institution Press.

Hogan, R. L. (2007). The historical development of program evaluation: Exploring the past and present. *Online Journal of Workforce Education and Development, II*(4). Retrieved from pdfs.semanticscholar.org/ee2f/dbbe116a30ab7a79b19e1033a7cab434feec.pdf

Hussey, M. A., & Hughes, J. P. (2007). Design and analysis of stepped wedge cluster randomized trials. *Contemporary Clinical Trials, 28*(2),182–191. doi:10.1016/j.cct.2006.05.007

Juras, R., Minzner, A., & Klerman, J. (2018, November 8). *Using behavioral insights to market a workplace safety program: Evidence from a multi-armed experiment.* Paper presented at the Fall Research Conference of the Association for Public Policy Analysis and Management (APPAM), Washington, DC.

Kershaw, D., & Fair, J. (1976). *The New Jersey income-maintenance experiment.* New York, NY: Academic Press.

Klerman, J. A., Wolf, A., Collins, A., Bell, S., & Briefel, R. (2017). The effects the Summer Electronic Benefits Transfer for Children demonstration has on children's food security. *Applied Economic Perspectives and Policy, 39*(3), 516–532. doi:10.1093/aepp/ppw030

Litwok, D., & Peck, L. R. (2018). Variance estimation in evaluations with no-shows: A comparison of methods. *American Journal of Evaluation, 40*(1), 104–118. doi:10.1177/1098214017749318

Manzi, J. (2012). *Uncontrolled: The surprising payoff of trial-and-error for business, politics, and society.* New York, NY: Basic Books.

Meyer, M. N. (2015). Two cheers for corporate experimentation: The A/B illusion and the virtues of data-driven innovation. *Colorado Technology Law Journal, 13*(2), 274–331.

Murphy, S. A. (2005). An experimental design for the development of adaptive treatment strategies. *Statistics in Medicine, 24*(10), 1455–1481.

Olsen, R. B., Bell, S. H., & Nichols, A. (2017). Using Preferred Applicant Random Assignment (PARA) to reduce randomization bias in randomized trials of discretionary programs. *Journal of Policy Analysis and Management, 37*(1), 167–180. doi:10.1002/pam.22005

Peck, L. R. (2015). Using impact evaluation tools to unpack the black box and learn what works. *Journal of MultiDisciplinary Evaluation, 11*(24), 54–67. Retrieved from http://journals.sfu.ca/jmde/index.php/jmde_1/article/view/415

Peck, L. R. (2016a). *Evaluation design for program improvement.* Paper presented at the Welfare Research and Evaluation Conference, Washington, DC.

Peck, L. R. (2016b). On the "how" of social experiments: Analytic strategies for getting inside the black box. *New Directions for Evaluation, 152,* 85–96. doi:10.1002/ev.20211

Peck, L. R. (Ed.). (2016c). Social experiments in practice: The what, why, when, where, and how of experimental design & analysis. *New Directions for Evaluation, 152.* San Francisco, CA: Jossey-Bass/Wiley. Retrieved from http://tinyurl.com/NDEonExperiments

Peck, L. R., & Scott, R. J., Jr. (2005). Can welfare case management increase employment? Evidence from a pilot program evaluation. *Policy Studies Journal, 33*(4), 509–533. doi:10.1111/j.1541-0072.2005.00131.x

Peck, L. R., Moulton, S., Gruenstein Bocian, D., DeMarco, D., & Fiore, N. (2019). *The first-time homebuyer education and counseling demonstration baseline report: Short-term impact report.* Washington, DC: U.S. Department of Housing and Urban Development, Office of Policy Development and Research. Retrieved from https://www.huduser.gov/portal/sites/default/files/pdf/Short-Term-Impact-Report.pdf

Peck, L. R., Werner, A., Harvill, E., Litwok, D., Moulton, S., Rulf Fountain, A., & Locke, G. (2018a). *Health Profession Opportunity Grants (HPOG 1.0) impact study interim report: Program implementation and short-term impacts* (OPRE Report #2018-16a). Washington, DC: Office of Planning, Research, and Evaluation, Administration for Children and Families, U.S. Department of Health and Human Services. Retrieved from https://www.acf.hhs.gov/opre/resource/health-profession-opportunity-grants-hpog-10-impact-study-interim-report-implementation-short-term-impacts

Peck, L. R., Werner, A., Rulf Fountain, A., Lewis Buell, J., Bell, S. H., Harvill, E., . . . Locke, G. (2014). *Health profession opportunity grants impact study design report* (OPRE Report #2014-62). Washington, DC: Office of Planning, Research and Evaluation, Administration for Children and Families, U.S. Department of Health and Human Services. Retrieved from https://www.acf.hhs.gov/sites/default/files/opre/hpog_impact_design_report_11_14_14_r2_0.pdf

Peck, L. R., Zeidenberg, M., Cho, S.-W., Litwok, D., Strawn, J., Sarna, M., . . . Schwartz, D. (2018b). *Career pathways design study: Evaluation design options report.* Washington, DC: U.S. Department of Labor. Chief Evaluation Office. Retrieved from https://www.dol.gov/asp/evaluation/completed-studies/Career-Pathways-Design-Study/5-Evaluation-Design-Options-Report.pdf

Puma, M., Bell, S., Cook, R., Heid, C., Broene, P., Jenkins, F., . . . Downer, J. (2012). *Third grade follow-up to the head start impact study final report* (OPRE Report #2012-45). Washington, DC: Office of Planning, Research and Evaluation, Administration for Children and Families, U.S. Department of Health and Human Services. Retrieved from https://www.acf.hhs.gov/sites/default/files/opre/head_start_report.pdf

Rossi, P. H. (1987). Introduction. In J. L. Miller & M. Lewis (Eds.), *Research in social problems and public policy: A research annual* (vol. 4). Bingley, England: Jai Press.

Rossi, P. H. (2003, October). The "Iron Law of Evaluation" reconsidered. Paper presented at the Fall Research Conference of the Association for Public Policy Analysis and Management (APPAM), Washington, DC.

Schochet, P. Z., Burghardt, J., & McConnell, S. (2008). Does Job Corps work? impact findings from the National Job Corps study. *American Economic Review, 98,* 1864–1886.

Shadish, W. R., Cook, T. D., & Campbell, D. T. (2002). *Experimental and quasi-experimental designs for generalized causal inference.* New York, NY: Houghton Mifflin.

Singal, A. G., Higgins, P. D. R., & Waljee, A. K. (2014). A primer on effectiveness and efficacy trials. *Clinical and Translational Gastroenterology, 5*(1), e45. doi:10.1038/ctg.2013.13

Slavin, R. E. (2013). *Why control groups are ethical and necessary* [Blog]. Huffpost.com. Retrieved from http://www.huffingtonpost.com/robert-e-slavin/why-control-groups-are-et_b_4114350.html

Stern, E., Stame, N., Mayne, J., Forss, K., Davies, R., & Befani, B. (2012). *Broadening the range of designs and methods for impact evaluations* (Working Paper #36). London, England: Department for International Development.

Thomke, S., & Manzi, J. (2014, December). Spotlight innovation on the fly: The discipline of business experimentation. *Harvard Business Review* No. R1412D. Boston, MA: Harvard Business School Publishing.

U.S. Department of Health and Human Services, Administration for Children and Families, Family and Youth Services Bureau. (n.d.a). *Transitional Living Program Evaluation.* Retrieved from https://www.acf.hhs.gov/fysb/long-term-outcomes-tlp

U.S. Department of Health and Human Services, Administration for Children and Families, Office of Planning, Research and Evaluation. (n.d.b). *Job search assistance evaluation, 2013–2018.* Retrieved from http://www.acf.hhs.gov/programs/opre/research/project/job-search-assistance-evaluation

U.S. Department of Housing and Urban Development, Office of Policy Development and Research. (n.d.). *The family options study.* Retrieved from https://www.huduser.gov/portal/family_options_study.html

Walker, J. T., Copson, E., de Sousa, T., McCall, T., & Santucci, A. (2019). *Pilot testing a randomized controlled trial of the transitional living program.* Washington, DC: U.S. Department of Health and Human Services, Family and Youth Services Bureau and Office of Planning, Research, and Evaluation.

Weiss, M. J., Bloom, H. S., & Brock, T. (2013). *A conceptual framework for studying the sources of variation in program effects* (Working paper). New York, NY: MDRC. Retrieved from http://www.mdrc.org/sites/default/files/a-conceptual_framework_for_studying_the_sources.pdf

W. K. Kellogg Foundation. (n.d.). *Logic model development guide.* Retrieved from https://www.wkkf.org/resource-directory/resource/2006/02/wk-kellogg-foundation-logic-model-development-guide

GLOSSARY

A/B testing (see *treatment vs. alternative treatment*)

activities—In a program logic model, activities refers to what a program offers and how it consumes or transforms inputs in order to generate program outputs

average treatment effect—the difference in mean treatment group outcomes and mean control group outcomes

blocking—a feature of random assignment that involves forcing assignment, so that, over some preset number of randomizations, the chosen ratio be met exactly

business as usual control group (see *control group, status quo, business as usual, existing service environment*)

causal model (see *logic model*)

competing treatments design—a multi-armed experiment that compares two (or more) alternative program models

conceptual framework (see *logic model*)

control group (no services)—represents a counterfactual where *no alternative services* are available, as might be the case in a new kind of program for which there are not existing alternatives

control group (status quo, business as usual, existing service environment)—represents a counterfactual where the control group still has access to any services that are available in the community

counterfactual—what would have happened and does happen in a program's absence

crossovers—control group members who receive the intervention "(in opposition to their experimental group assignment)"

differential attrition—a threat to internal validity; different rates of follow-up data collection for treatment and control groups, resulting in nonequivalent groups; also referred to as experimental mortality

efficacy trials—evaluations of interventions under ideal, controlled circumstances

effectiveness trials—evaluations of interventions under real-world conditions

enhanced treatment design (scale up, scale back variants)—scaling up (or scaling back) the features of a program in order to compare the base version of an intervention to an enhanced (or scaled back) version of that program. This approach allows measuring the contribution of the added component, for example, to the intervention's impact magnitude

evidence review (see *systematic reviews*)

external validity—the extent to which the results of an evaluation are generalizable to other settings, populations or times; a synonym is generalizability *factorial design (full factorial, fractional)*—varies two (or more) treatment dimensions or factors, randomizing to each individually and to both together. If the levels of each factor include "absence" or "presence," then the absence of both factors represents a status quo control group

false negative (Type II error)—missing the detection of a favorable impact

false positive (Type I error)—falsely claiming a favorable impact exists

fidelity—the extent to which the program model in theory is fully implemented in practice; a synonym is implementation fidelity

fractional factorial design (see factorial design)

historical forces (or history)—a threat to internal validity; historical, social, political or economic events, trends or forces that influence outcomes of interest

impact—the change in an outcome that is attributable to the program

impact evaluation—isolates the change in outcomes that the program caused from the many other possible explanations that exist

implementation evaluation—documents the implementation or operations of a program, including determining whether the program activities have been implemented as intended and possible reasons for divergence; synonyms are operations research, process evaluation

implementation fidelity (see fidelity)

implementation research (see implementation evaluation)

input—the leftmost part of a standard logic model that identifies what resources (e.g., human capital, financial, legislative, etc.) go into a program

instrumentation bias—a threat to internal validity; the influence of changing measures or data collection procedures on outcomes

intent-to-treat impact (ITT)—an impact estimate that involves comparing the whole treatment group to the whole control group, regardless whether any of the treatment group members were "no-shows," and regardless whether any of the control group members finagled their way into treatment and were "crossovers"

internal validity—an evaluation design's ability to support causal claims

intervention logic (see logic model)

large-scale experiments—usually consider broad and long-term implications of policy change; and, as such, take a fair amount of time to plan, implement and generate results

logic model—in order for a program to articulate how it expects to achieve change, the four elements of the logic model must be identified and their relationships stated; synonyms or alternative names include: conceptual framework, program logic, program logic model, program flow, program theory, theory of change, causal model, results chain, intervention logic

maturation—a threat to internal validity; the influence of people or institutions' natural evolution of outcomes

mediator—the variable (such as program take-up, dosage or quality) through which the effect of treatment occurs

meta-analysis (meta-evaluation)—involves quantitatively aggregating other evaluation results in order to ascertain, across studies, the extent and magnitude of program impacts observed in the existing literature

meta-evaluation (see meta-analysis)

minimum detectable effect (MDE)—The MDE is the smallest true intervention impact that can be detected under the given empirical circumstances, such as the evaluation's sample size, data and measures, and statistical assumptions

moderator—distinct group defined by baseline traits

multi-armed experimental evaluation—an evaluation design that includes three or more groups that are randomized to, and may or may not include an excluded control group

multistage experimental design (see staged evaluation design)

natural experiments—a class of evaluations that involves allocating access to or exclusion from a treatment group by natural forces, those not manipulated by a researcher but that emulate random assignment

no-shows—treatment group members who do not receive the intervention

nudge experiment—an experiment that focuses on behavioral insights or administrative systems changes that can be randomized in order to improve program efficiency

null results—failure to detect statistically significant results

opportunistic experiment—an experiment that takes advantage of a given opportunity: when a program has plans to change—for funding or administrative reasons—the evaluation can plan an experiment to coincide with that change

outcome—the construct that a given program aims to achieve either as the ultimate goal of the program or as step towards the program's ultimate goal

outcome contrast (see impact)

plausible rival explanation—a logical alternative to the research hypothesis being examined

pretest-posttest evaluation design—an evaluation design that involves comparing one group's posttest outcomes to the same group's pretest outcomes to estimate a program's impacts

process evaluation (see implementation evaluation)

program flow (see logic model)

program logic (see logic model)

program logic model (see logic model)

program theory (see logic model)

random assignment (randomization)—the process of using a random process for assigning units—be they individuals, classes, schools, organizations, cities, and so on—to gain access (or not) to an intervention involves something akin to a coin toss or a roll of the dice

random assignment ratio—the number of those randomized into treatment relative to the number of those randomized out of treatment (or into an alternative treatment)

rapid-cycle evaluation—this evaluation type aims to produce results quickly and provide decision makers with timely and actionable evidence of whether operational changes improve program outcomes

regression artifacts—a threat to internal validity; that program targets are chosen for being extreme in some way, either at the top or the bottom of a distribution, and are therefore likely to return or "regress" to the mean naturally, regardless of the intervention or program; a synonym is regression-to-the mean

regression toward the mean (see *regression artifacts*)

results chain (see *logic model*)

selection bias (two variants: creaming, attrition/differential attrition)—a threat to internal validity; people who choose to or who are chosen to participate in a program differ from others in ways that influence their outcomes

sequential, multiple assignment, randomized trial (SMART; see staged evaluation design, (adaptive treatment)

staged evaluation design (adaptive treatment)—this staged evaluation design involves randomly assigning units a second time into a second-stage treatment that is determined by their initial response to treatment; a synonym is multi-stage experimental design; a type is sequential, multiple assignment, randomized trial (SMART)

staged evaluation design (trigger response)—this staged evaluation design involves randomly assigning units a second time into a second-stage treatment if they reach a certain trigger point; a synonym is multi-stage experimental design

staggered introduction evaluation design—a design that involves staggered, sequential roll-out of an intervention, essentially creating a "waiting list" at random, giving all participants, as clustered into groups, access over time

statistical power—statistical power is equal to the probability of rejecting the null hypothesis if the alternative hypothesis is true (or, one minus the probability of a "false negative" result, or Type II error). In other words, power relates to the analyst's ability to detect an impact, should it exist.

statistical significance—the statistical significance level is the probability of identifying a "false positive" result (also referred to as Type I error)

status quo control group (see *control group*)

stepped-wedge evaluation design—this design staggers access to the intervention randomly so that initially some units/groups serve as controls and eventually the entire pool of eligibles gains access

stratification—in randomizing units within an experiment, groups (or strata) can be identified and used either to ensure population representation or to establish over-representation. There might be research or administrative reasons to give a greater (or lesser) chance to some categories of people to be selected to treatment

subgroup analysis, split sample—analysis of subgroup impacts that involves dividing the sample (both treatment and control groups) into distinct groups as defined by their baseline traits (e.g., men in one group, and women in another) and then executing the impact analysis on each group

subgroup analysis, interacted model—analysis of subgroup impacts where the sample is kept intact and an "interaction" term is added to the estimating equation to identify impacts by subgroup

systematic reviews—a type of literature review that uses systematic methods to collect secondary data, critically appraise research studies, and synthesize studies; a synonym is evidence reviews or tiered-evidence reviews

testing bias—a threat to internal validity; the influence of the act of being tested on outcomes

theory of change—a fully elaborated program logic model where not only the inputs, activities, outputs and outcomes are identified, but also external influences (and plausible rival explanations for changes outcomes) are represented; a frequent (though imprecise) synonym is logic model

threats to internal validity—factors external to an evaluation design that compromise its ability to establish causal claims

tiered-evidence reviews (see *systematic reviews*)

treatment contrast—the difference between the treatment and control conditions

treatment group—group that receives treatment in an experiment

treatment on the treated (TOT)—the impact for those who took up the treatment offer, as opposed to the effect of being offered it

treatment vs. alternative treatment—variant of the basic, two-armed experimental evaluation design that compares an alternative treatment to an existing program, sometimes through comparing alternative models, sometimes through adding a specific enhancement, and other times through scaling up or back; a synonym is A/B testing; types include competing treatments and enhanced treatment designs

Type I error (see *false positive*)

Type II error (see *false negative*)

variance of the impact estimate—a measure of the "noisiness" of the impact estimate; the variance of the impact estimate is a function of sample size, with larger samples producing lower variance and therefore more precise impact estimates

INDEX

Made in the USA
Las Vegas, NV
29 August 2021